The Leadership Imperative

The Leadership Imperative: Technology Adoption and Strategic Management in Travel Firms in Jamaica

By

Andrew Spencer

CAMBRIDGE SCHOLARS PUBLISHING

The Leadership Imperative:
Technology Adoption and Strategic Management in Travel Firms in Jamaica,
by Andrew Spencer

This book first published 2013

Cambridge Scholars Publishing

12 Back Chapman Street, Newcastle upon Tyne, NE6 2XX, UK

British Library Cataloguing in Publication Data
A catalogue record for this book is available from the British Library

ISBN (10): 1-4438-4249-4, ISBN (13): 978-1-4438-4249-5

I dedicate this work to my darling wife and soul mate, Diana. You have always supported my dreams and believed in me in ways that even I don't believe. You allowed me to go thousands of miles away from you to follow my dreams and then gave up your life to follow me half way across the world. Sharing this journey with you daily has made the book worth doing and only confirms that you're the meaning and the inspiration in my life. I love you with all my heart.

TABLE OF CONTENTS

LIST OF ILLUSTRATIONS

LIST OF TABLES

PREFACE

This work is the product of research for a PhD in Tourism Strategy conducted at Bournemouth University. The motivation came from observations of the fragile nature of the Jamaican retail travel market in an age of growing e-tourism and technology adoption. Classical theories of adoption such as the diffusion of innovation theory, and the technology acceptance model, and literature on pertinent theories which have been highlighted as drivers of adoption such as the *Resource-Based View, Firm Strategy, Culture* and the *Digital Divide* have been thoroughly reviewed. Prior to this however, the afore-mentioned classical adoption theories are contrasted with the Post-Internet debate which explored *Information Asymetry* and *Disintermediation*. Having conducted this review it was determined that the leadership/ownership role had not been sufficiently emphasized in technology adoption, therefore this PhD thesis sought to more clearly identify these personal factors in combination with the previously explored factors. The overarching theory of *Organizational Decision-Making* was used to provide a framework to identify drivers of decision-making processes in general and then apply these to the internet adoption context. This resulting book aims to identify the combination of antecedents of technology adoption for travel firms and distil factors to identify the key determinant of the adoption of the internet for sales and marketing purposes in small, owner-managed travel firms. It examines the firm characteristics which are associated with adoption behaviour such as strategy and resources, as well as external factors such as culture and the digital divide. In addition to external and firm factors, personal factors such as ownership and leadership are explored at various stages of adoption.

A predominantly qualitative methodology was used to interview travel agencies in the context of Jamaica. All firms which have similar characteristics in terms of ownership and management structure, in particular where owners are themselves the managers and provide leadership for the organization, were interviewed. The owner-managers of these firms were interviewed to gather deep perspectives from local industry experts on industry challenges, current technology involvement and future directions. Exploratory descriptive quantitative methods were used to analyze firm characteristics and their relationships to internet adoption for sales and marketing as well as the intention to use these technologies in firms, while a deeper exploration into owner-managers was achieved through qualitative enquiry. The findings indicate that the leadership role is more significant than has been previously posited.

The contribution to knowledge is new in that it takes a unique approach to an understanding of technology adoption in firms by creating a comprehensive conceptual framework for adoption based on previous research and then creates a model that shows the factors and variables that drive adoption at each stage of the adoption process from a personal leadership perspective as well as the organizational perspective. Ultimately it is hoped that this focus on each stage of adoption will provide insights into firm adoption behaviour as a consequence of leadership characteristics.

ACKNOWLEDGEMENTS

I must first say thanks to my supervisors, Professor Dimitrios Buhalis and Dr. Miguel Moital for detailed guidance and continuous support throughout this journey. Their expert advice and rigourous assessment have shaped this book into what it is today. Heartfelt gratitude to Dr. Dalea Bean, who has been supportive before day one and who has committed a significant amount of time to editing, proofreading and giving general advice throughout the process.

A special thanks to other individuals in the School of Tourism at Bournemouth University who influenced my journey. Thanks to Professor Alan Fyall for having the confidence in me and providing the positive outlook on all situations and to Dr. Michelle McLeod for providing useful tips along the way. The other PhD candidates in the school have also helped to enrich the experience through informal discussions, social gatherings and just a genuine interest in the progress of my studies.

The Department of Management Studies at the University of the West Indies has provided financial support which helped to make the arrival at this destination possible. Special mention must be made of Professor Gordon Shirley, Dr. Densil Williams and Dr. Noel Cowell for ensuring that this support was provided. The encouragement to take on this challenge was provided by Dr. Anne Crick and Mr. Stanford Moore, and to them I am grateful for the gentle push to get an arduous process started.

Also at the University of the West Indies, Professor Anthony Clayton is credited with providing the start to the critical analysis of concepts and theories which were pertinent to this study. His constantly thorough examination of my work was responsible for enhancing the quality of my written work and the connecting logic in each argument. Thank you Prof. for believing that I was "PhD material".

My sincerest appreciation goes out to my parents James and Eunice for giving me the positive start that I needed and for not giving up on me along the way when a University education was far from my mind. Your belief in my abilities and your desire for me to excel were the driving force for all of my achievements. Even though we were thousands of miles away, you always ensured that you were aware of every pitfall of this journey through constant calling and caring. I love you both. To all my friends and family at SDCOG (Reformed) in Jamaica and the UK, who kept me in your thoughts and your prayers, a big thank you is extended to you and I will be forever indebted. Most of all, thanks to God for the inspiration and guidance from beginning to end.

INTRODUCTION

This work focuses on the adoption of Information and Communication Technologies (ICTs) in firms by moving from the generic to the specific. Mainstream theories of Strategic Management and Organizational Decision-Making provide useful insights about organizational behaviour in general, and inform the more focused discussion surrounding technology adoption decision-making and in particular technology adoption decisions in small, owner-managed travel firms.

The idea of using ICTs in the travel industry is not a novel one and stems from the first reservation systems in the 1950s to the tourist information systems like TIS and Gulliver of the 1980s (Werthner 1995(a); Werthner 1996; Werthner and Klein 1999) to major global distribution systems (GDSs) like SABRE and Amadeus in the 1990s to the enormous number of current travel activities on the Web. Now more than ever however, it brings a multiplicity of players together with relatively easy access to each other and each other's information base. The implication of this is clearly that a more level playing field has been created with more options available to the creators of products and services and the final consumer. Internet booking brings new independence for suppliers and travelers (Poon, 2001). New value chains and value systems are emerging and the players within the industry have to redesign their strategy under the power of ICTs (Buhalis, 2002).

While this work ultimately focuses on the adoption of the internet for sales and marketing purposes, there are certain preconditions which must be met in order for firms to reach certain levels of ecommerce adoption. The work of Moital et al. (2009) espouses the concept of innovation interdependence which identifies the necessity of previous technology conditions in order for ecommerce to take place. For example, the adoption of the internet is necessary in order for internet sales and marketing activity to take place. Likewise the adoption of computer technology is important in firms in order for internet adoption to take place.

This dependency relationship at each level of adoption allows for this work to be approached from the innovation interdependence perspective and facilitates the exploration of all levels of technology adoption (computers, internet and ecommerce). The logic is that the understanding of a higher level of technology adoption presupposes an understanding of lower levels of adoption which are prerequisites to that higher-order adoption practice. Therefore the hierarchical approach of investigation in this study assesses computer adoption, internet adoption, website adoption (general information), ecommerce adoption and social media adoption.

Since 2000 it has been argued that there are varying levels of internet readiness on the international landscape. Canadians and North Americans in general were keen users. According to Law and Leung (2000) Europe lagged behind North America by 14-18 months for internet penetration. Although there is now a smaller disparity between Europe and North America, the developing world and in particular the Caribbean typically lags behind both. There are obvious implications for countries which are slower on the uptake of ICTs and for businesses operating in those contexts. Intermediaries in Jamaica, which are predominantly travel agencies that focus on the outbound traveler, may not experience some global issues such as disintermediation at the same rate as more developed economies and may therefore become less strategic and proactive. This is due in part to feelings that the locals who are outbound travelers in the region are not internet enthusiasts, because of a high risk perception in making online purchases; a view which coincides with the work of Lin et al. (2009).

The travel firms in Jamaica are predominantly outbound agencies; however a discussion of the inbound travel climate and the digital divide (unequal technology access) which exists between tourists and destinations is useful to the extent that inbound travel provides a gap which may be filled by these local agencies. The inbound market has been largely ignored by these firms on the grounds that it does not provide a commission in the same way that outbound travel does; this represents a short-sighted view and does not take into consideration issues of value-adding and dynamic packaging. While the digital divide may have been reduced, more recently, Minghetti and Buhalis (2010) still identified that there are multiple technological divides which exist between tourists and destinations within developed countries and between developed and developing countries. This will lead to varying levels of digital exclusion.

Previous research on the digital divide has not explored the tourism context in the way that this has been done by the work of Minghetti and Buhalis (2010). In particular it points to important marketing and communication challenges between tourism generating countries and tourism destinations. It is increasingly evident that high-tech tourists, regions or enterprises meet in an electronic marketplace and interact directly on electronic platforms, which greatly reduces the need for spatial as well as temporal synchrony for information and distribution. Minghetti and Buhalis (2010, p. 278) articulate that:

> "The study of the digital divide is critical for less technologically developed regions that need to expand their ICT usage to be able to promote their offerings, interact with consumers, and reduce their dependency on intermediaries."

An important argument from this article is that medium- and low-digital-access tourists and destinations still depend on analog transactions and physical intermediaries to develop their planning processes for stimulating vacations in these destinations. This is very applicable to the Jamaican context which is a relatively low –digital access destination catering to high-digital access markets such as the United States and the United Kingdom. There is still a fairly high dependence on physical intermediaries in these tourist generating countries as the tourist board and tourism and hospitality enterprises maintain sophisticated relationships with agencies in the American Society of Travel Agents and the Association of British Travel Agents.

Three issues emerge as a result of this arguable overdependence. The first is that high-tech tourists in these high-access countries who prefer to interact in the electronic marketplace may not be captured in the island's marketing efforts. The second is that the destination image may be affected when high-tech tourists are unable to sufficiently interact with the destination in cyber space. According to Govers et al. (2007, p. 19) "covertly induced and autonomous agents, in particular, have a dramatic influence" over destination image in the minds of consumers. These agents include television, magazines and the internet. While a significant amount of Jamaica's marketing budget is spent on television advertising (Williams and Spencer, 2010), internet promotion and interaction receive far less attention. The major implication is that initial television exposure may lead potential tourists who are high-tech, in search of information and booking options on the web, which may create a poor destination image when these needs are not met.

The third and most significant issue is that local travel agencies in Jamaica do not play a role in driving inbound travel and tourism to the island. These agencies, which have seen a decrease of more than 50% in the number of companies in operation over the last ten years, typically cater to the outbound market. This is so because they do not get a commission for inbound travel sales. The ability of agents to leverage their importance may however come from an improved capacity to meet the need for an online presence for the country. These agencies now receive little attention from statutory bodies and a greater contribution to the island's tourism may present a strong case for governmental collaboration and support. The challenge however is that many of the firms do not engage in more than simple internet activities of emailing, while less than 5% of these firms have active company websites, according to preliminary data in this study. The question of what stimulates these companies to adopt various levels of internet technology becomes a pivotal one to any discussion of their changing role.

Industry Challenges and the External Environment

This work was influenced by the observation that travel agencies in Jamaica, particularly at the time when the concept of travel disintermediation was fairly new to the region, started facing significant challenges with drastic commission reductions that saw the number of firms decline from 105 to 43 over the period of 1999-2009, according to the president of the Jamaica Association of travel Agents (JATA). A closer examination revealed that there were very few changes to operational procedures in these firms over this period despite the changes which were taking place in the global market place. The most static area of operations related to technology use. Subsequent to the adoption of Global Distribution Systems (primarily Sabre and Amadeus) in the 1990s, there was very little new adoption of technologies in particular those which have a direct impact on sales and marketing despite the global internet revolution.

This presented the need to understand why there was such a low level of adoption in these firms despite the need for a rejuvenation of business practices through innovative business models. An understanding of why internet adoption for sales and marketing is low in these firms may provide insights into an explanation for why firms with similar characteristics in similar contexts have high or low levels of adoption. This research takes a comprehensive look at possible drivers before identifying a key determinant. The theoretical context further explains pertinent issues and challenges in the global environment.

Studies related to technology adoption may be categorized as pre-internet or post-internet phase studies. Dominant works in the pre-internet phase had their foundations in the work of Rogers (1962) and Davis (1989). Fundamental differences in these schools of thought lie in their approaches to understanding the drivers of adoption. In Rogers's (1962) diffusion of innovations, he focuses on innovation, communication and the role of the social system, while in Davis's (1989) technology acceptance model, the emphasis is on individual perceptions about whether a particular technology innovation is easy to use or meets individual needs (usefulness).

In the post-internet phase, the debate suggests that technology is now more pervasive and widespread than earlier technologies such as those which enhanced the productivity of farming processes for example those originally studied by Rogers (1962). The industrialized world over the last 30 years has been making a transformation from an industrial economy to an information economy, and this may be seen in the development of the literature. According to Parker (1988), information, rather than land or capital, will drive the creation of wealth and prosperity over the next several decades. Technology is irreversibly changing the business world and internal organizational operations. Drucker (1990) further argues that there is a transformation in which it is knowledge and not capital, natural resources or labour, which has the greatest impact as a means of production.

Technology which facilitates knowledge transfer allows for competitive advantage for businesses (Porter, 2001). There is a difficulty however in providing a succinct definition of ICT. Buhalis (2002) points out that ICTs include hardware, software, groupware, netware, as well as the intellectual capacity to develop, programme and maintain the equipment. Due to the complexity and breadth of these electronic tools that facilitate the strategic management of organizations, it is important to note that this book aims to focus on what is arguably the most pervasive of these tools; the internet.

Although conceived in 1969 the internet never had widespread use as we know it today until 1991. In comparison to other technologies the internet spread much faster throughout the world and all dimensions of organizations and industries have to be re-examined in the light of the power of this new ICT (Klein 1996; Bakos 1998; Amit and Zott 2001). Other theorists in the *post-internet phase* argue that the internet affects every part of the business's operation from internal processes to external relationships, as well as modifying and restructuring entire economic sectors (Kalakoa and Whinston 1996; Gatty 1998; Ghosh 1998; Timmers 1998; Wirtz 2001). The reality is that the reach of the internet has spread across geographical regions more than any other technological development (Tang and Yang, 2011). The most striking characteristic is its ability to permeate multiple economic sectors and industries and transform the business operations internally and externally.

The Research Context

Travel and tourism has emerged as the leading economic activity due to the economic crisis of 2008 which has affected the bauxite industry in Jamaica. Understanding the local economic environment within which these firms operate is paramount. The country is a middle-income, oil-importing country that attempted diverse economic development strategies during the 1970s and 1980s. The major sectors of the economy were bauxite and alumina, tourism, manufacturing, and agriculture (Statistical Institute of Jamaica, 2010). Bauxite and alumina, in particular, set the pace for Jamaica's post World War II economic growth through new investment and foreign exchange earnings. Bauxite production declined rapidly in Jamaica in the 1980s however, because of the prolonged recession in the world aluminum industry, global oversupply, and the departure of multinational producers. Tourism declined in the 1970s, but recovered between 1980 and 1986, thus becoming the second most important sector of the economy. According to Meditz and Hanratty (1987), manufacturing, a quite diversified sector, underwent structural changes in the 1980s when production was refocused on exports rather than on the domestic market. Agriculture, the heart of the Jamaican economy for centuries, has been in relative decline for decades since World War II.

With a population of 2.7 million people (STATIN, 2010), the social context has played a role in shaping technology perceptions as well as leadership traits in this relationship-oriented society. A major vehicle for this influence is the formal education system. Before independence from Great Britain in 1962, much of the educational system was provided by churches. Since then, the state has brought most schools into the public sector, although some private schools remained (Whyte, 1983). The authorities have prioritized universal literacy and, in the early twenty-first century, 88 percent of adult Jamaicans could read and write. Tuition is free in so-called basic (primary) and secondary schools. Compulsory education lasts, in theory, until age 16, but some 120,000 children aged between 12 and 16 are not enrolled in school. Higher education courses are offered at a campus of the University of West Indies in Kingston as well as at University of Technology (formerly the College of Arts, Science and Technology), Northern Caribbean University, the Norman Manley School of Law, the College of Agriculture, the United Theological College, and teacher training colleges. In recent years numerous offshore universities from the United States have added to the number of offerings.

Studies in tourism have existed at the University of the West Indies since the 1970s, with students completing their final 2 years in the Bahamas. In 2006 the campus in Jamaica created a full three year programme for students wishing to stay in the country to complete their tourism education. The other universities mentioned above have also contributed to tourism and hospitality training. Technology training saw significant growth when the College of Arts, Science and Technology became the University of Technology. These studies are usually undertaken by the more scientifically inclined and the trickle down of technology skills to the rest of the country has been slow; a direct impact on business in the island.

The travel firms operate within the limitations of the technology which is available in the society which makes the technological context a critical element of understanding firm behaviour. Despite other challenges with broadband access for the population, Jamaica has a fully digital telephone communication system with a mobile penetration of over 95%.The country's three mobile operators: Cable and Wireless (marketed as LIME – Landline, Internet, Mobile and Entertainment), Digicel, and Oceanic Digital (operating as MiPhone and now known as Claro since late 2008) – have spent millions in network upgrade and expansion. Both Digicel and Oceanic Digital were granted licences in 2001 to operate mobile services in the newly liberalised telecom market that had once been the sole domain of the incumbent Cable and Wireless monopoly.

With wireless usage increasing, landlines supplied by Cable and Wireless have declined from just over half a million to roughly about three hundred thousand as of 2006. A new entrant to the Jamaican communications market, Flow Jamaica, recently laid a new submarine cable connecting Jamaica to the United States. This new cable increases the total number of submarine cables connecting Jamaica to the rest of the world to four. The implication is that a desire to engage in ecommerce could be supported by this improvement in technology infrastructure.

As a developing nation, Jamaica has much to gain from being a full participant in the global economy. Because the country's primary trading partner is the United States, it is critical that they establish and maintain the infrastructure necessary to engage those consumers as fully as other nations around the world. Given the limited GDP of the nation and the skills of its inhabitants, Jamaica must continue to develop or face record deficits in Balance of Payments. Tourism, Agriculture, and Mining currently provide a substantial portion of the country's revenue. All of these local industries suffer from deficient technology by U.S. standards.

Jamaica has never been a major tourist generating country. Most of the travel associated with the island is inbound. Tourism began in Jamaica in the 1890s, when the United Fruit Company, seeking to use the excess capacity of its ships, encouraged cruises to Jamaica, and tourist hotels were constructed on the island. Tourism, however, did not flourish until after World War II, when accelerated depreciation allowances for investment in that sector helped to triple the number of hotels between 1945 and 1970. Inbound travel to the island has traditionally been dominated by the US traveler due to close proximity. Outbound travel has also largely been to the United States due to a high concentration of the Diaspora in that country. As a major destination Jamaica receives far more tourists than it generates. Data on the travel activities of Jamaicans to other countries for vacation or business is virtually non-existent; however it is expected that locals typically travel to visit friends and relatives given the limited economic means of the average household. While the actual statistics do not exist for outbound vacation travelers, the number of Jamaicans living in other countries suggests that there is a fairly significant pull factor for locals to visit.

The Jamaican Diaspora is unusually large, with some estimates indicating that as many individuals of Jamaican descent may currently be living outside the country as within it. The Diaspora is concentrated in three countries: the United States, Canada, and the United Kingdom. About 637,000 Jamaican foreign born lived in the United States in 2008, with approximately 123,500 in Canada in 2006 and 150,000 in the United Kingdom in 2008. This data is according to the official statistical bureaus in each country. Travel from Jamaica is usually handled by physical intermediaries. These outbound travel agencies have faced significant challenges over the last decade. This is evidenced by an alarming decrease in the number of agencies with the number moving from 105 to 43 between 1999 and 2009. In 2000 American Airlines, British Airways and the regional carrier Air Jamaica cut commission from 9% to 6% for travel agencies in the region. According to the *Jamaica Gleaner* (2000) when Jamaican travel agencies attempted to boycott the sale of American Airlines tickets in response to the 3% cut in commissions in 2000, their efforts proved futile as the airline, which accounts for approximately 70% of air traffic to the region, simply decided that direct bookings was an option. The agencies' position was further weakened when Air Jamaica and British Airways followed suit with similar cuts just two months later. Further commission cuts were experienced in 2009 from 6% to 3%. This and other changes in airline reservation models has undoubtedly lead to displacement of some Jamaican travel agencies, even resulting in some unregistered agencies transacting business illegally. According to the Minister of Tourism, these agencies have no insurance or protection for the client and several have got into trouble with customers who pay for travel which did not materialise (*Jamaica Gleaner*, 2011).The article also states that in an attempt to punish these rogue agencies, the Jamaican parliament has recently approved the legislation to increase the fine from J$20 for each day they operate illegally, to a maximum fine of J $1million.

It is clear that the current business model where these firms only cater to the outbound market to provide a limited set of services seems to be failing. This research posits that in addition to a re-examination of the market, the business practices and processes within these firms such as their technology usage rate, types of technology adopted and the purposes for which they have been adopted, must be assessed with a view to informing a more adaptable business model.

A Note on Methodology

The aim of the research investigation is to identify and assess the key driver/barrier of internet adoption for sales and marketing purposes in small, owner-managed travel firms. Consequently the following objectives emerged:

1. To examine the combination of antecedents and drivers for various levels of technology adoption in travel firms.
2. To investigate the relationship between Ownership/Leadership and technology adoption in owner-managed small firms.
3. To investigate the influence of internal firm factors such as strategy and resources in technology adoption in owner-managed small firms.
4. To investigate external firm factors such as the digital divide and culture in technology adoption in owner-managed small firms.
5. To develop a model of technology adoption for owner-managed small travel firms that identifies the characteristics of leaders and firms at varying levels of technology adoption.

While the debate has existed mainly between Positivism and Interpretivism, and while these have both influenced the methods employed in numerous research projects, there is greater grounding in the research perspective surrounding *Critical Social Science* to achieve the aims of this work. Embedded in *Critical Social Science* is the idea that there is an opportunity to provide critique that reveals true conditions and help people into action

(Neuman, 2006). It focuses on supplying individuals and societies with the tools needed to change the world around them. More importantly, it allows for exploration through multiple layers beneath the surface, to uncover underlying structures and highlights that the purpose of social science is to understand social meaning in context. A dialectical orientation is adopted as knowledge allows people to see and alter deeper structures. Fundamentally however, the primary reason for research under this paradigm is that it aims to smash myths and empower people to change society (Neuman, 2006).

This validates the need to explore the Jamaican context to identify the multiplicity of considerations facing local agencies, as opposed to adopting a "one-size fits all" approach that would immediately suggest that the drivers of diffusion of innovation and technology acceptance are universal. It also supports the idea that the firms should never be comfortable in existing situations but attempt to explore all possibilities, to include threats and opportunities. The research aims to address the issues through deep explorations into the thoughts and feelings of stakeholders to gain explanatory insights. This research process is primarily an inductive one with a deductive element, as it explores issues through the deep qualitative inquiry of a small sample and identifies what is generalizable.

If assessed on a continuum between *Positivism* and *Interpretivism*, this research finds more of its philosophical underpinnings in *Critical Social Science*. More specifically within this paradigm, there is a subscription to the concept of *Bounded Autonomy,* which is an approach to human agency and causality that makes the assumption that there is some degree of subjectivity in human actions but only within limits that are clearly identifiable. Therefore it recognizes that there is a certain degree of objectivity in those boundaries or limits but that within those parameters there may be room for subjectivity in exploring causality and the actions of the human agent. This approach therefore blends determinism (a feature of positivism) and voluntarism (a feature of interpretivism) to show how structure and agency interact. Previous research on technology adoption has primarily utilized deterministic, positivist approaches (see for example Davis, 1989; Bagozzi et al. 1992; Lederer et al. 1998; Fuchs et al. 2009). The blended approach in this work operates on the foundation that people will make choices but only based on what they deem to be possible, which is a function of what they think can happen within identifiable limits. These limits to subjectivity may be cultural or material. Cultural factors are those constraints placed on an individual's autonomy such as values, beliefs and norms; while material factors are those which surround resource constraints. With this in mind the research attempts to explore the reality of Jamaican travel agencies, which may or may not be significantly different from other realities. Jamaican travel agencies have been chosen due to a significant decline in the number of firms and the lack of research in the travel industry since its inception. The lessons learned here may bear some relevance to an understanding of similar industries and societies. The essential characteristics to determine similarity include small, owner-managed travel firms in relationship-oriented, developing economies.

Since this research is primarily a qualitative one, which adopts an inductive approach, it does not warrant the extensive use of existing measurements in the planning phase as this would likely cripple the flexibility needed to engage in the Template Analysis in an effective way. Neuman (2006) argues that qualitative researchers typically do not refine abstract ideas into theoretical definitions early in the process. Instead rudimentary working ideas are refined during the data collection and analysis process. Through gathering and analyzing qualitative data, the researcher develops new concepts, formulates definitions for major constructs and considers relationships among them. These are eventually linked to create theoretical relationships. Conceptualization therefore flows from the data which has been collected. Operationalization in this process comes before conceptualization. It however still takes place during data collection and analysis. The operationalization of qualitative research is a description of how working ideas were developed while making observations and collecting data. It is more of an "after the fact" description than a preplanned technique which shows how specific data and preliminary ideas about this data become constructs.

The concepts of reliability and validity are typically notions within the positivist paradigm. While reliability addresses the extent to which results are consistent over time or generalizable, validity is concerned with the extent to which a study measures that which it intended to (Golafshani, 2003). Whereas these two issues may never be fully resolved in the qualitative paradigm, there is still a need to demonstrate that research in this paradigm is credible and trustworthy. According to Patton (2002) validity determines reliability in qualitative research. The research design while using dominant qualitative methods, incorporated closed ended quantitative data which was analysed in tandem with qualitative data to make the analysis more robust. Although the quantitative, deductive element of the study was minimal it provided a foundation for deeper qualitative inquiry. Using the semi-structured interview, the series of questions were guided by the objectives and a rudimentary conceptual framework. This provided a deductive element, however the dominant inductive approach has been validated through a detailed template analysis technique to identify themes across cases. The results are generalizable and transferable only to the extent that samples share the characteristics of small, owner-managed travel firms in a similar developing country context.

While it may be argued that the level of reliability and validity is reduced through adopting a qualitative approach, the dominant approaches taken to the study of technology adoption, leadership, culture, digital divide, resources and strategy have been predominantly quantitative and reductionist. A critically important observation is that the two major concepts being researched (technology adoption and leadership) suffer from similar methodological constraints. The majority of studies on technology adoption rely heavily on survey-based quantitative approaches (see for example Bagozzi, Davis and Warshaw, 1992; Lederer, Maupin, Sena and Zhuang, 1998; Fuchs, Hopken, Foger and Kunz, 2009). The same is true for mainstream leadership studies (see for example Cho, 1998; Cole and Mehran, 1998; Anderson and Reeb, 2003; Welch, 2003; Ghobadian and O'Regan, 2006). These studies have been important in identifying key variables for further exploration but the intellectual debate is now at a juncture where it would benefit from deeper insights which cannot be gained from factor analysis or ordinary least squares regression.

The Primary Research Process

The nature of the work necessitated a pilot study and two phases of data collection. The pilot study tested the instrument for phase 1 which included all of the relevant constructs identified in the literature such as culture, the digital divide, resources, strategy and leadership. Ten respondents were interviewed to determine the suitability of questions.

The data collection for phase 1 was aimed at meeting the following objectives:

- To investigate the relationship between Ownership/Leadership and technology adoption in owner-managed small firms.

- To investigate the influence of internal firm factors such as strategy and resources in technology adoption in owner-managed small firms.

- To investigate external firm factors such as the digital divide and culture in technology adoption in owner-managed small firms.

The above objectives comprise three of the five objectives in this study and are output related. In this book there are four output objectives and one input/process-related objective. This means that four of the objectives are expected to be met through the findings of the primary research while one objective is aimed at influencing the data collection itself. The input/process-related objective emerged from the literature and helped to design the instrument which was used in phase 1 of data collection. This objective was:

- To examine the combination of antecedents and drivers for various levels of technology adoption in travel firms.

This instrument was constructed by utilizing concepts which were tested in a variety of studies related to technology adoption. In order to probe and ask the most appropriate questions a number of concept maps in addition to variables and constructs used in previous research, are used to create a guide. These concepts emerge from the rudimentary conceptual framework seen in *Fig 1-2.* While conceptual frameworks do not typically emerge this early in qualitative research, a flexible, revisable, rudimentary framework helps in streamlining ideas for the qualitative researcher (Vaughan, 2008). The interview instrument was a semi-structured one. The unstructured approach was not taken because the concepts are not completely new but are rather being applied in a new way. Despite a dominant qualitative methodology being used, the first set of questions on technology adoption used a closed-ended approach for the phase 1 instrument to allow for a descriptive analysis of the respondents' technology involvement. The subsequent sections of the instrument which focus on organizational decision-making, leadership and macro-factors, are designed predominantly with open-ended questions which allow for deep explorations into attitudinal concerns and processes within the firm.

Theorists in organizational decision-making theory such as Huber and McDaniel (1986) highlighted that organizational environments of the future are likely to be characterized by greater complexity. They called for organizations to recognize that the centrality of the organization rests in what decisions are made and how these decisions are made. This makes it particularly important to get to the root of what drives decisions for individuals and ultimately organizations. Given the complexity mentioned above a single component model of attitude would not be suited for the instrument. The decision was therefore taken to use affective-cognitive models of attitude otherwise known as the multi-component model. This has primarily been used in psychology and consumer behaviour research (see for example Schiffman and Kanuk, 2000; Youn, 2000) however it has been applied to the organizational decision-making framework in this research. It has also not been used in the study of firm technology adoption prior to this work.

Rogers (1994) innovations adoption model provides a comprehensive group of innovation characteristics which have been adapted and applied to assess leaders' perceptions of an innovation. Given the fact that the leaders being interviewed are also owners of these firms, three key characteristics were explored initially as these directly affect an owner's return on investment: relative advantage, perceived risk and image. Compatibility and Complexity were later assessed based on how employees react to technologies.

- *Relative advantage* was addressed using five statements which addressed the general importance of technology in work processes and moved to more detailed questions about making work simpler and faster (Eason, 1988).
- *Perceived risk* associated with using technology was addressed in three open-ended questions which addressed the risk of investment from the perspective of time and money (Wernerfelt, 1984; Mowery et al. 1998) given that owner-managers are likely to focus on these types of returns on investment.
- *Image* was measured through two questions: one relating to how competitors view the firm and the other related to customers' views (Gronroos, 1993). These constructs were taken from the work of Gronroos in the marketing literature.
- The issue of complexity was covered through questions about whether technology was easy to use (Davis, 1989), easy to learn to use (Lederer, 1998) and whether it made work easier or harder.
- *Compatibility* was assessed by whether a technology was a good fit for the tasks being carried out in organizations (Peteraf, 1993). One open-ended question borrowed from the work of Petaraf covered this issue. Additionally the researcher developed a question which asked how the benefits of a new technology were assessed before being introduced.

The primary data in this phase was collected using semi-structured in-depth interviews with the top executive of 31 travel agencies in Jamaica. The total population of agencies is 43, however only those with owner-managers were interviewed as their level of autonomy in firms is greater.

Additionally only those firms which were International Air Transport Association (IATA) certified were interviewed. The other firms were either a part of chain operations or did not have IATA status. The entire target sample of 31 was interviewed as there were no major limitations to collecting this data apart from the time constraint for executives. The questions in the interview brief were guided by the conceptual framework and the literature; however participants were allowed to speak freely about other related matters. The instrument employed closed-ended questions for firm description followed by probing open-ended questions. Interviews were taped and transcribed to control interview length and ensure that all data was recorded.

Interviews were taped and transcribed. The decision was taken to manually identify themes through complete immersion in the data. This process involved writing a case study using all the data for each firm and then conducting a cross-case and comparative analysis across responses to questions in the semi-structured interviews through the use of an outcomes matrix. This allowed for the grouping of answers according to themes. Common themes were then identified among firms which had similar types of adoption levels. From the phase one analysis adoption levels were identified as computer adopters, internet adopters, website adopters, ecommerce adopters and social media adopters. This was determined based on similarity in technology use among groups of firms.

Computer adopters was used to describe firms at the bottom of the hierarchy that were simply engaged in the use of computer terminals and hardware for back office accounting functions or for front office functions such as sales. Firms which do not use online sales tools other than GDSs are also placed in this category since their adoption was simply based on the free provision of the system by the supplier. *Internet adopters* represent firms in which the internet is used only for emailing and web browsing. *Website adopters* refer to the firms which have created and used company websites for general and marketing information sharing. Websites in this case typically provide static information. *E-commerce adopters* refer to firms which use websites for actual bookings and payment, while *social media adopters* would refer to firms in which social media is used for promotion and interaction which results in sales activities. Each firm was categorized as being at one of two stages within each of these levels based on the work of Damanpour (1991): initiation or implementation. While the data analysis for phase one allowed for description of the firms, it did not allow for the question of "why" to be answered sufficiently. After completing this phase, the question of why firms were at each level could not be answered conclusively, as the factors of digital divide, culture, strategy and resources, did not adequately answer this question as discussed in the subsequent findings chapters. Phase one however highlighted an importance in the leadership role. Although this indicated some degree of causation, it did not allow for the development of a model, which could explain each level of adoption which emerged. At this stage it was determined that a second phase of data collection would allow for more robust findings, which could identify leadership variables that may influence adoption behavior in firms. This was guided by constructs in the most dominant leadership research surrounding the transformational/transactional paradigm.

The exploration of owners' feelings through in-depth, semi-structured interviews allowed for the adoption of the Template Analysis technique (Crabtree and Millar, 1999). Template analysis is the process of organising and analysing textual data according to themes. This can be text produced or used in the context of the evaluation irrespective of the evaluation activity, i.e. data not generated by the evaluation (Crabtree and Miller, 1999). This therefore allows for broader underlying structures to be included in the analysis. This was also strengthened through a cross-case analysis matrix (Patton, 2002). The in-depth interviews were transcribed verbatim so that the exact words of respondents could be used in the categorizations to follow.

Phase 2 was designed to meet the following objective:

- To develop a model of staged technology adoption for owner-managed small travel firms that identifies the characteristics of leaders and firms at varying levels of technology adoption.

In meeting this objective the second instrument was designed in a much more focused manner than the previous instrument and relied solely on an open-ended qualitative-type instrument to gather data. The instrument used in phase 2 is a completely qualitative one, which seeks to strengthen and validate the leadership constructs and variables identified in phase 1 of the primary data collection. These questions were more grounded in the transformational leadership literature, as leadership type was identified as the major influencing factor in driving various levels of adoption for the firms under investigation. The instrument is a very concise one which seeks to be extremely focused and not as broad and comprehensive as the previous instrument. This instrument which focuses on transformational leadership attributes used the 4 general constructs identified by Bass and Aviolo (2003) in the Multifactor Leadership questionnaire as a guide.

These constructs are *idealized influence, inspirational motivation, intellectual stimulation and individualized consideration.* The first few questions were however used to clarify issues related to technology experience, risk, and education; all of which emerged as significant themes in phase 1 of the study and needed greater depth. The 31 owner-manager firms which were examined in phase 1 were contacted for a second round of interviews. Only 20 of these firms expressed an interest in being interviewed another time. The major reason given by these top executives was the time constraint. This number of firms was substantial enough for the nature of this qualitative study and significant data was gathered about the leadership styles and its influence on technology adoption levels. Through a second phase of interviews which focused on transformational leadership constructs, a relationship was identified between transformational traits and firm adoption levels. Additionally leaders who possessed more transformational characteristics had certain descriptive elements in common such as education level and type, previous work experience, family composition, technology experience, risk aversion and intellectual stimulation. These similarities enabled the formation of leadership typologies at each level of technology adoption within the firms.

While education level was determined from re-visiting the data from phase one, the remaining variables of education type, risk aversion, family composition, previous work experience, technology experience and intellectual stimulation were determined from this second phase. Intellectual stimulation was the only construct from the Multifactor Leadership Questionnaire constructs (Bass, 1999) which showed differences in technology adoption behavior for leaders; however when combined with the other variables there was remarkable explanatory potential. The template analysis technique, which was used in phase 1, was also employed here along with a cross-case analysis matrix (Patton, 2002). The fundamental difference was that this phase did not have any closed-ended questions. The results led to the generation of descriptive summaries which facilitated explanations of particular variables and led to an analysis and evaluation on how gaps in the literature may be filled. A combination of both phases of data collection and analysis allowed for the development of case studies. The approach taken to the data analysis allowed for single cases to be written and then compared for similarities. This facilitated the inclusion of case studies at each adoption extreme to be presented to highlight contrasts between the characteristics of leaders at both ends of the hierarchy of adoption.

Limitations

This book provides for greater depth through the lens of *Critical Social Science* by giving credit to previously unearthed variables in the predominantly positivist studies of the last few decades, while at the same time using these constructs and variables to probe into latent constructs. In doing so the book adds greater depth to technology adoption and leadership while adding greater breadth to areas such as the resource based view and strategic management. Nonetheless, one major challenge faced in collecting data included the time constraint of owner-managers. This was due especially to their integral involvement in the daily operations of the business. Most interviews therefore had to take place in the evenings when business activity was reduced. One result of this was tiredness in each interview, which typically lasted for approximately an hour. Additionally some respondents were concerned about being tape-recorded but they were assured of anonymity. The anonymity of respondents was maintained throughout the discussion of the findings based on signed agreements with respondents that their identity would be protected in the research process. The analysis of the data provided its own challenges in terms of identifying patterns in the qualitative data and doing the cross-case analysis. However, full immersion in the qualitative data as well as support from some exploratory quantitative descriptions, assisted in the emergence of patterns and typologies (See *Fig. 5-7*).

The Structure of the Work

Chapter One provides a review of the theories surrounding technology adoption decision-making, which encapsulates the focus of this work. In doing so it addresses strategic management, organisational decision-making and technology adoption processes. This chapter also addresses antecedents in the technology adoption discourse such as the digital divide, culture, leadership and resources. The chapter concludes with two conceptual frameworks showing how the constructs from the literature interact.

Chapter Two characterises the firms interviewed based on exploratory closed-ended questions, which highlight the current technology activities of the firm and the intention to use other technologies. This data was primarily gathered from closed-ended questions. This chapter also presents data collected through open-ended questions about the role of strategy and resources in making decisions about whether to adopt technologies in the firm. Issues of risk are also addressed here.

Chapter Three looks at macro factors, such as culture and the digital divide which drive adoption decisions such as norms, values, traditions and technology access. It focuses on how these external factors influence innovative behaviour as well as the making of leaders in firms.

Chapter Four is a pivotal chapter which emphasizes the role of leadership in technology adoption as the key determinant. While other factors play a role, the data reveals that leadership occupies the most central role and the discussion of these findings are elaborated in this chapter.

Chapter Five gives a breakdown of how the emergent model from this research was developed through various stages of the process. It begins with a re-assessment of the earlier conceptual framework in chapter one and presents a revised iteration before demonstrating how single components of the final model were identified and how each was connected to the other to provide a coherent model.

The Conclusion provides a final analysis of how this work meets the research objectives and contributes to the body of knowledge. It highlights why this research is new and why it is worthy of a doctoral designation by showing its intended impact on theory and practice.

CHAPTER ONE

FIRM LEADERSHIP AND TECHNOLOGY ACCEPTANCE: RETHINKING ADOPTION PERSPECTIVES

Strategic management theory addresses the long term vision of organizations. It also speaks to competitiveness and factors that drive competiveness. The literature on strategic management provides scope for one of the objectives in this study which is to investigate the role of internal firm factors such as strategy and resources in technology adoption in owner-managed small firms. The competitiveness approach was pioneered by Porter (Porter 1985; 2001) which states that a company develops its business strategies to obtain competitive advantage over its competitors. In his seminal work Porter (1985; 2001) points to five competitive forces, which have been extensively used by researchers (David, 2007; Kim et al. 2004; Stonehouse and Snowdon, 2007; Poon, 1993; Wynne et al. 2001) over the last three decades.

Much of the research however has indicated that firm strategy must be influenced by competitors as well as markets. While these are useful arguments they are both output driven discussions with very little emphasis on inputs such as ownership, leadership and resources. It is critical that this work fills the gap in discussing input considerations. Also, while previous research has focused on markets and competitors, it has only superficially assessed the impact of strategy on technology adoption and other processes. Competitive strategies therefore need to become a pivotal part of the adoption discourse.

In assessing the five forces model discussed by Porter (1985) in conjunction with discussions by David (2007), there seems to be consistency surrounding the issue of competitiveness. The literature in both cases speaks to the need for special attention to be given to the external environment as well as internal capabilities of organizations to survive in the same environment. As previously articulated by Porter (Porter 1985; 2001) a company develops its business strategies to obtain competitive advantage over its competitors. It does this by responding to five primary forces:

- The threat of new entrants
- Rivalry among existing firms within an industry
- Threat of substitute products/services
- Power of suppliers
- Power of buyers

Porter's model on the five competitive forces in an industry addresses some key concerns for companies operating in a globalized world. He describes five competitive forces that a company has to deal with effectively to achieve a competitive advantage in the business environment. He suggests that an effective strategy must take into account not only the actions and reactions of direct rivals, but also the roles of suppliers and customers, alternative products that satisfy the same basic need, and the prospect that new entrants will enter the market (Porter 1985). This model focuses on the value chains within an industry, which involves looking beyond the boundaries of a company. It describes a network of players that perform related value creation activities and that have to be considered in the strategy development for a company. Adapting Porter's model to an eBusiness strategy can provide a useful framework for e-commerce decisions since it considers internal capabilities as well as the external environment.

The work of Enright and Newton (2005) addresses a previous gap in the work of Porter's proponents by asserting that the importance of competitiveness attributes may vary across locations, depending on product mix and target market segments, especially in complex, multifaceted industries such as tourism. A dominant argument from Porter's five forces model which is still resonant is that companies remain competitive by continually adapting to changes in the external trends and internal capabilities, competencies and resources regardless of the industry (David, 2007). He continues by saying companies must effectively formulate, implement and evaluate strategies that capitalize on those factors. It seems to be clear from a strategic perspective therefore that where the external environment changes, companies must have the ability to keep pace and stay ahead in a competitive environment. It therefore cannot be business as usual for organizations that hope to compete. Competitive strategies influence other strategies within firms such as strategies for distribution of products and services.

Christopher (1992, p. 4) in his early work on distribution strategies defines distribution as "the process of strategically managing the movement and storage of materials, parts, and finished inventory from suppliers through the firm and onto its customers." He further discusses that distribution strategies involve the much wider consideration of logistics related to inventory levels, materials management and information systems as with transport. While there was an awareness of the need for competitive distribution strategies, there were still major challenges in the implementation of such strategies because there was the need for new management skills (De Hayles and Taylor, 1972; Persson, 1978; Christopher, 1986). These skills involve complete systems management, customer service management and operations coordination.

The literature on distribution strategies shifted focus from internal management of the organization to a greater emphasis on channel strategy. Gattorna (1994), states that the most strategically significant challenge facing organizations is the development of a channel strategy. While much of the literature in previous decades (Chandler, 1963; De Hayles and Taylor, 1972; Persson, 1978; Christopher, 1986) focused on physical distribution of tangible products, Gattorna highlights that any progressive organization including service organizations such as banks, airlines and insurance companies should attempt to select the right strategy for getting the final product or service to the consumer. However his definition of distribution channels still reflects a preoccupation with tangible goods. He defines the distribution channel as the commercial arrangement established to enable a product to flow from the point of production to the point of ultimate consumption. Even with a focus on tangible distribution it was identified here that in some cases a change of companies may be needed and in others it may be suitable. In order for a distribution channel structure to be established, authors (Rushton and Oxley, 1993; Gattorna, 1994; Hatton, 1994; Chorn, 1994) argue that three determinants emerge. They identify the first as being the requirements of the final customer as measured by an aggregation of customers with similar requirements. The channel strategy will therefore depend on the constituents of the segment and what route will reach them best. This therefore means that e-commerce adoption may be determined by whether the market segment is predisposed to online or offline transactions.

The second determinant is identified as the capabilities of the organization, which supports the resource-based perspective that says that a firm's strategic decisions are in large part influenced by its resources. Capabilities may also refer to management capabilities which were also highlighted in the literature on ownership and leadership. The third determinant is stated as the availability and willingness of intermediaries (if needed) to participate in the channel. The discussion of intermediaries has been limited to their role as a conduit for the supplier; very little is however said about the determinants of their own distribution strategies.

Discussions of channel strategy however, moved to a broader debate of managing the entire supply chain. Particularly within the discourse on international distribution (Rushton and Oxley, 1993) it has been posited that there needs to be the careful management of all companies involved in the distribution of goods and services. The focus must be on the fact that the end result must be one in which the target market is reached efficiently (Chorn, 1994). This indicates an emphasis on the needs of the final consumer and also the need for the supplier to use resources in an efficient, cost-effective manner. The literature is however limited in that it argues that the intermediary must be managed by the supplier who needs to control these variables. It neglects to discuss that the intermediary must also mange the relationship in two directions within the supply chain.

Schary and Larsen (1998) state that the concept of the supply chain embraces a number of elements. Firstly it identifies the complete process of providing goods and services to the final user and brings the actions of supplier and customer into a single system. This is coordinated through an information system accessible to all members. They highlight that the main objective of the supply chain is service to customers; however this must be balanced against costs and assets. Supply chain members however do not do this on their own, as individual achievement is dependent on the performance of the chain as a whole. It is the management of a network of interconnected businesses involved in the ultimate provision of product and service packages required by end customers (Harland, 1996). All activities must be coordinated well together in order to achieve the least total logistics cost. The travel supply chain involves the supplier, wholesaler, retailer and consumer. The supplier here represents those creating and providing the service such as airlines, hotels, attractions and car rental companies. The wholesaler refers to tour operators such as Thomas Cook and Thomson Travel who create packages while retailers are travel agencies which sell directly to customers in the marketplace. It is clear then that as the customer contact agent an exploration is needed to determine what factors drive or deter these travel agents' engagement in e-commerce.

Hamel and Prahalad (1994) made the observation that global distribution becomes the most crucial element in a global strategy, the conduit for products to reach the market and the guardian of global market share. Suppliers attempting to compete internationally must manage their distribution effectively. This effective management will involve claiming market shares while managing cost and satisfying market share in order to retain and maintain control. Travel agents must then position themselves competitively while meeting the competing needs of the supplier and consumer.

Establishing a presence in particular markets affects the bottom line of organizations. Suppliers however must do cost benefit analyses to ensure that benefits in dollars outweigh the cost of gaining market access and distributing products. Many firms such as Google do this through strategies that enable them to give away the product free of charge. Terpstra (1991) adds from a marketing perspective that there are three distribution tasks: gaining entry into a foreign market, serving many markets at the same time and establishing presence as an insider to each market. Supply chains can be manipulated and modified to suit the needs of the supplier and customer. The ultimate goals of competitiveness and seeking a sustainable and defensible competitive advantage have become the concern of every manager who is alert to the realities of the marketplace. It is not acceptable to assume that good products will sell themselves, neither is it advisable to imagine that success today will carry forward into tomorrow (Christopher, 1998). On this basis organizations are seeking for avenues through which to meet the needs of markets while minimizing their distribution costs. With contemporary theorists (Woodward, 2000; Smith, 2007) arguing that distribution involves the dissemination of information, the means of booking and purchasing, product bundling and packaging, researchers should no longer merely focus on the distribution strategies of suppliers but pay more keen attention to other firms such as intermediaries who must also make critical decisions about how they reach their markets and transact business.

Reduced transaction cost and commission present a strong case for complete elimination of travel intermediaries (Buhalis and Licata, 2002). When Caribbean travel agencies attempted to boycott the sale of American Airlines tickets in response to the 3% cut in commissions in 2000, their efforts proved futile as the airline, which accounts for approximately 70% of air traffic to the region, simply made it known that "direct bookings" was always an option. The agencies' position was further weakened when Air Jamaica followed suit with similar cuts just two months later. Hatton (2004) states that the reduction of commissions paid and the competition provided by the Internet present uncertain operating environments. In 2003, British Airways reduced agents' income in the UK to 1%. What is clear from these events is that organizations must do what is necessary to ensure profitability; this may either be done by finding new ways to increase revenues or by reducing costs, or a combination of both. With inflation, competition and rising oil prices, revenue increases in many cases will only be marginal as cheaper prices may not be passed on to the consumers. Due to high fuel costs and competition, the heavy focus is therefore on reducing cost especially those related to distribution.

Pearce (2009) argues that suppliers and their location destinations must seek to develop more effective distribution strategies. He further suggests that the tourism marketing literature has taken a tier-by-tier approach rather than a network approach to understanding distribution design, which has led to channel width considerations being poorly addressed. A network approach however, while useful in drawing attention to relationships between many players, does not address the deeper operational concerns of specific players. In cases where the research on tourism distribution channels does this, it tends to focus on either suppliers or consumers. Alternatively the literature (Crotts et al. 1998; Karamustafa 2000) gives attention to relationships between wholesalers and suppliers or between intermediaries (Lumsdon and Swift, 1999).

This work contends that numerous relationships have been explored but there needs to be greater emphasis on the determinants of a travel retailer's decision to choose particular business strategies. There has been much work on travel agencies (Palmer and McCole, 1999; Barrett and Standing, 2001; Dale, 2003; Law et al. 2004) however this work has largely focused on general approaches which must be taken by agencies to avert disintermediation, one of which has been identified as internet adoption. While these arguments are compelling, many agencies still do not engage in this practice, therefore deep qualitative research is needed to understand the related causes. This work attempts to do this, by applying generic theories of the firm such as Resource Based Perspectives, Ownership, Leadership, and Strategic Management.

Intermediaries do not operate at either end of the supply chain therefore a look at their buying relationship with suppliers is a critical element of the debate surrounding the strategies which they employ. The concept of strategic purchasing is not a new one and theorists (Eames and Norkus, 1988) have long discussed relationships in business to business (B2B) transactions. They have in fact suggested that business buyers should create a strategy that consists of four elements: self-analysis, vendor-analysis, vendor-performance analysis and programme review. The purchaser must conduct detailed research before entering into an agreement. The implication here is that buyers must adopt a systematic approach to transacting with vendors. This is vital as the stock or input purchased from another business will affect the final output of the buying company.

Relational competence has been paramount in the discussions surrounding strategic procurement management and addresses the ability to manage relationships with other firms in the supply chain (Cox, 1998). He articulates that an effective business strategy is one in which firms recognize that their boundaries need to change constantly in response to consumer preferences. This theoretical position built on the work of many theorists (Ricketts, 1994; Williamson, 1979; 1990; Reve, 1990) which focused primarily on reactive and simplistic approaches to purchasing and supply management. Most of this previous analysis addressed the interplay between asset specificity and transaction cost, which presupposes that a firm's operating structure, will be fixed. The issue of asset specificity in particular was treated poorly in the literature, as companies such as intermediaries source externally regardless of whether services have low or high asset specificity. Due to the nature of their operations they have little room for sourcing internally.

Cox (1998) therefore extends the discussion to address shifting boundaries and what he calls "strategic supplier alliances" where single sourced relationships may emerge with a supplier of a complementary product or service. These usually involve a much closer relationship between the main contracting firm and the complementary supplier and are often referred to as joint ventures. The systems used by the primary supplier may therefore influence the systems used by the complementary supplier. In the case of an intermediary who relies solely on a supplier for stock, this influence is likely to be more pronounced. For example if the primary supplier only sells stock electronically, secondary or complementary suppliers are somewhat coerced into implementing systems that facilitate their purchase of this stock, especially where these relationships discussed by Cox are manifested. He therefore convincingly argues for a relational competence analysis, which involves a determination of what the efficient boundaries of the firm are so that these can be created to reduce transaction costs and improve quality and value whether collaboratively or competitively.

Virolainen (1998) however later argued that the literature on the strategic management of procurement has focused on particular products but less so on the procurement function and he therefore describes these approaches as narrow. He therefore proposes an integrated procurement strategy which involves value-chain positioning, objectives setting, organization structuring, strategic make or outsource decisions, choice of different strategies, company level strategy and functional strategies.

While Virolainen has attempted to move from the specific nature of other studies to a more general framework, ostensibly what it is needed is even greater specificity as the nature of procurement varies based on industry and relationship. Additionally there has been little done in this general framework to assess the nuances in procuring a product as opposed to a service.

More recently however a continuation of the discussion of relationships has been carried out by Ryals and Humphries (2007), who state that while unnoticed to business marketers, there has been a quiet revolution in supply chain management where the traditional emphasis on least-cost transactions has given way to a focus on long-term relationships with a few key suppliers. This again is in contrast to the previous views of some authors (Ricketts, 1994; Williamson, 1979; 1990; Reve, 1990) and demonstrates a shift from the view that transaction cost is the key driver. This provides further reinforcement of the previously articulated point that these long-term relationships are likely to lead to a relationship between primary suppliers and complementary suppliers or intermediaries where processes and practices of the primary may filter through to these secondary operators or dependents. Procurement Strategies have focused in large part on offline modes of distribution, however competitive environments demand that some companies shift focus to an exploration of e-procurement given that supply chain-related activities accounted for the spending of 70% of a firm's revenues (Presutti, 2003). A discussion of ecommerce activities is critical therefore as it looks at online transactions in both directions along the supply chain; e-procurement which focuses on business-to-business activities and e-distribution which highlights business-to-customer transactions. These have implications for the overall debate on firm strategy.

ICT can change the structure of an industry and alter rules of competition (Porter and Millar, 1985; Porter, 2001). They also contend that ICT can be used to create sustainable competitive advantage and provide companies with new competitive instruments. As a result of ICT, new business can be developed within a company's existing activities. The emphasis of research on the relationship between strategy and ICT in the strategic management literature tends to focus on how ICT influences competitive, distribution and procurement strategies. The symbiotic relationship is largely ignored and recently Beckinsale et al. (2011), while recognizing the important role of ICT in strategy, have emphasized the role of strategy in ICT adoption. This re-conceptualization and reassessment of the causal direction is critical as previous research has been preoccupied with how ICT modifies strategy and industries but in large part has failed to demonstrate how different types of strategies affects levels of adoption for technology-based innovations.

Porter (2001) suggests that technological change, such as ICT, is one of the most important forces that can alter the rule of competition. This is because a majority of activities in an organization create and use information and this will be more powerful for industries where information is the key product. Until recently travel agents have operated with privileged information as their primary stock in trade (Cheyne et al. 2006). They argue that the Internet is a major source of information and so the agencies' stock is at risk due to the availability of information to end users in an unprecedented manner. They now need to find new ways to gain and maintain a competitive advantage. David (2007) states that strategic management is all about gaining and maintaining competitive advantage. He argues that competitive advantage is essential for the long term success of any organization or industry. Firms can only sustain this for a certain period of time due to rival firms imitating or undermining that advantage. It is therefore not sufficient to simply gain competitive advantage but to also achieve a sustained competitive advantage for the economic sustainability of industries and organizations. Firm strategy discourse must however consider how strategy decisions are made within firms. Strategy decisions fall within the scope of theories of organizational decision-making, which will also provide a framework within which to discuss technology adoption decision models.

Organizational Decision Making (Over-arching Theories)

The adoption of any innovation in organizations speaks to the broader issue of how the decision process unfolds in organizations. Organizational decision-making theory therefore provides an overarching framework for the analysis which will be carried out. There are two main models in organizational decision-making; models which emphasize the individual decision-maker at a micro level and those which address macro decision making at the organizational level. The assumptions of the micro models rest on the notion that there is a cerebral rationality of each decision-maker which overrides situational influences in decision-making. The contrasting macro models highlight the influence of organizational context on the decision-making process.

The cerebral rationality concept, a micro model, was first espoused by Simon (1957). Researchers such as Conlisk (1996), Rodan and Galunic (2004) and Mercer (2005) recognized its importance in psychology, economics and international politics and support the notion that good decision-making emerges from rational thought processes of the decision-maker. A resonant and recurring ideological discussion supports the idea that understanding the individual decision process is the key to disaggregating the organizational decision process. The emphasis is therefore on each decision-maker's performance. The contrasting approach of macro decision-making theorists builds on the work of Cyert and March (1963). In attempting to broaden the scope of the discourse, Langley et al. (1995) highlighted that the future of organizational decision-making research must address more complex issues such as the fact that the decision-maker is not only driven by cerebral rationality but also by experiences. Additionally single decisions should be viewed as a network of many issues, which points to the evidence of dynamic linkages rather than simple linear ones. More recently Kwon et al. (2009) advocated the need to integrate both approaches and assess micro decision-making within macro decision-making parameters. With this in mind this work attempts to identify the multiplicity of concepts which have been identified as pertinent to the study of the adoption of technology-based innovations by looking at how these concepts are interwoven through the broad framework of organizational decision-making.

Hubert and McDaniel (1986) highlighted that organizational environments of the future are likely to be characterized by greater complexity and therefore called for organizations to recognize that the centrality of the organization rests in what decisions are made and how these decisions are made. It is important therefore to get to the root of what drives decisions for individuals and ultimately organizations. Langley et al. (1995) emphasize the key role of individuals in organizations, who are decision-makers. This work informs the theoretical analysis especially because it analyses general typologies which have been ascribed such as:

- Decision-maker as *creator*- an intuitive individual who drives organizational decisions through creative insight.
- Decision-maker as *actor*- an individual who passively acts in accordance with what happens to the organization such as problems or opportunities.
- Decision-maker as *carrier*- an individual who carries with them their experiences and the impact of the world around them.

The seminal literature (Simon, 1957; Cyert and March, 1963) has treated the decision maker as more of a receptacle and as such treated the individual as an *actor*. Far less emphasis was then placed on the vision and abilities of the decision-maker. However more contemporary literature (Hitt et al., 2001; Ghobadian and O'Regan, 2006; Falk, 2008) has identified the importance of the insights of decision makers and these individuals are therefore treated as *creators*. The least emphasized in the organizational decision-making literature have been those categorized as *carriers*. This may probably be explained through the fact that both *creators* and *actors* would naturally have previous experiences based on the local and global context. This is however accounted for in the literature on culture and the digital divide, which has been assessed in this work.

Decision-makers in organizations are often referred to as leaders (Victorino et al. 2006). Therefore an important development in the organizational decision-making literature came from an assessment of leadership styles and decision-making. Tatum et al. (2003) identified that the long-standing distinction between transformational and transactional leadership had an impact on decision-making styles. Kuhnert and Lewis (1987) define a transformational leader as a visionary who is charismatic and influences strategic transformation through motivation. The transactional leader is defined as an efficient manager who focuses on the task at hand and uses a reward-driven approach. It would appear that these styles affect whether a leader adopts a restricted or comprehensive decision-making style (Tatum et al. 2003). This work therefore seeks to assess the leadership role in the decision-making by analyzing these typologies within situational contexts.

While the organizational decision-making theory highlights the role of the decision-maker, there has also been some focus on what Langley et al. (1995) refers to as *pooled linkages* and *contextual linkages*. *Pooled linkages* emphasize that decision-making issues are linked because they compete for financial resources and managerial time and energy. As such this research explores these resource issues by taking a deeper look at the *Resource Based View of the firm*. Alternatively *contextual linkages* address the role of broader issues such as culture, ideology, structure and strategy on the decision-making of the individual and organization. A deeper exploration of these issues will be examined through literature on culture and the digital divide. The approach of conducting the analysis through the lens of *Critical Social Science* is appropriate as this book attempts to assess the phenomena subjectively while maintaining that there are objective limitations; a feature of *the bounded autonomy concept*. The conceptual framework which encapsulates these ideas may be seen at the end of the literature review. The discussion however begins with a critical review of the literature on technology adoption models and more specifically internet adoption for sales and marketing, which is the behaviour being studied. The decision to adopt technology-based innovations in small firms is a very calculated one in light of limited resources. This therefore means that the decision-making process involved in adopting such new tools and processes must be assessed in trying to understand the choices which are made. In some cases decision-making processes may be complex and require many parties to approve a new innovation adoption, while in other small firms it may be a unilateral process which relies solely on the judgment of a key decision-maker.

Technology Diffusion and Adoption

The concepts of diffusion and adoption are two semantically different but relevant concepts in this discussion. Adoption refers to making use of an innovation within operational practices and processes while diffusion addresses how information about the innovation is transmitted in a group (Rogers, 1994). This work contends that these two concepts are very interrelated and symbiotic as knowledge transmission may affect the decision to use (adoption) and inversely adoption will necessitate diffusion. The diffusion and adoption paradigm saw growth in the 1940s with research from Ryan and Gross (1943). Almost two decades later the research in this area was limited due to the insularity of the research to the specific field. The applicability of the output to other fields was therefore very constrained. In 1962 Rogers created a transferable generic theory which allowed for this paradigm to be accepted across disciplines. This therefore led to the theory of diffusion and adoption of innovations being applied extensively. According to Moital et al. (2009, p. 52) his contribution to the development of diffusion and adoption theory is so substantial that his model "still prevails as the main theoretical source in the study of diffusion and adoption of innovations."

The work of Everett Rogers (1962; 1976; 1983; 1994; and 2003) has provided the foundation for much of the discussion on innovation diffusion. In his text *Diffusion of Innovations* he grapples with why obviously advantageous ideas take such a long time to be adopted. In other words even when an innovation will clearly bring significant benefits there is still a relatively long time before it is widely accepted or utilized. Diffusion is defined by Rogers (2003) as the process in which an innovation is communicated through certain channels over time among the members of a social system. The key variable is that new ideas are being communicated. He therefore describes the four key elements of diffusion as (1) innovation (2) communication channels (3) time and (4) social system.

Rogers (2003) makes a clear distinction between innovation and technology and refers to innovation as any perceived new idea, practice or object. While this includes technology it is not limited to technological advancement. The critical variable here is "newness" of anything which is being introduced. In many cases technology is a fairly new introduction into organizations and Latzer (2009) has taken a look at the possibilities of technological innovations being disruptive or sustainable as opposed to the earlier approaches of simply identifying these innovations as positive (see for example Bagozzi et al. 1992; Lederer et al. 1998). Keller (2008) sums the discussion up very well in stating that no single factor will determine the success or failure of technological innovation, but that it is the combination of many factors across disciplines such as social, political and economic, which may explain innovation effectiveness.

Communication channel speaks to how this "new" idea is exchanged or passed on from one individual to others (Rogers, 1994). This is also referred to by Loudon and Della Bitta, (1993) as the process of diffusion where knowledge of the innovation is transferred throughout a group of people. While diffusion of the innovation is critical to the adoption process it does not guarantee innovation adoption. It is however a prerequisite for adoption since the use of an innovation would require previous knowledge of its existence and usefulness. Hoffman and Roman (1984) viewed information flow as having two key dimensions in organizations: the origin of the information and the emphasis on innovation. The authors added profoundly to the knowledge of how innovations are diffused or communicated throughout organizations. Although leadership is not explicitly stated it is inherent in their argument. The second dimension in particular is dependent on leadership decisions and may be influenced through organizational culture or strategy. Additionally innovations are likely to be deemed more legitimate if they originate with organizational leadership. The authors should have more clearly emphasized the leadership role, which is what this work aims to do. Information flow about innovation may have become easier however through the speed of the internet, which has had a revolutionary impact (Xue, 2005). While this may hold partial validity, it does not explain what drives the diffusion of information about the internet as an innovation itself.

Of particular interest is Rogers's representation of a social system as a set of interrelated units that are engaged in joint problem solving to accomplish a common goal. With this definition he goes on to refer to units as organizations or informal groups. What is particularly useful is that he states that a system may represent actors in a village, community or country as well. He claims that social structure and norms of the system are therefore important. Rogers (2003) recognizes the importance of context even though much of his work has been generalized by authors. Yun and Avvari (2011) highlighted gaps in this research by emphasizing the role of innovation actors such as leaders in the public and private sectors. The contribution of their research is to place greater importance on the leadership element in innovation diffusion.

Rogers's work (2003) supports the leadership literature by highlighting that opinion leadership may be quite influential. He identifies that opinion leaders may act as change agents and assist with the innovation diffusion process. He then denotes that types of innovation-decisions are optional, collective or authority driven. This generalization for example is quite useful as contemporary observation of service firms reveals that *optional innovation-decisions* very rarely result in comprehensive adoption for the social system. *Collective innovation-decisions* usually gain more traction but are largely leadership driven even though there is consensus by the team. This consensus usually stems from a charismatic leader (Hitt et al., 2001; Bass and Avolio, 2003) or an opinion leader (Rogers, 2003). The reality is that while *authority innovation-decisions* may yield compliance, those with power may also be opposed to the diffusion of a particular innovation. This again magnifies the importance of ownership and leadership in the innovation diffusion discourse. It is then fair to say that while collective and authority innovation-decisions are more common in many organizations, these processes are however leader-initiated, which calls for an emphasis on the orientation of leaders and decision-makers.

Brown (1981) makes a distinction between various perspectives on innovation diffusion. He first addresses *The Economic History Perspective* which states that innovations are more likely to be adopted if they are more adaptable to the market in terms of value for money. He then highlights *The Adoption Perspective* which was pioneered by Rogers (1962) and states that while it is the traditional approach to diffusion studies, it is the most dominant and developed research perspective on the subject matter. In contrast, this perspective focuses heavily on innovation adoption as the outcome of a learning or communication process. He then analyses *The Market and Infrastructure Perspective*, another contrasting view which seeks to move away from the individual approach of the adoption perspective which presupposes equality for individuals and states that opportunities are "purposely unequal." It therefore focuses on whether surrounding conditions facilitate innovation diffusion. *The Development*

Perspective however extends *The Market and Infrastructure Perspective* to identify the impact of innovation diffusion on a society's welfare. It is a contrasting view to the previous three perspectives as it challenges the presupposition that innovation diffusion automatically has positive effects. Brown uses the example of third world nations where technological innovation has not led to significant economic development but rather served to increase disparities and reinforce elitism.

Each of the perspectives addressed by Brown (1981) provides useful insights into adoption practices, however the most applicable ones for contemporary analysis appear to be *The Adoption Perspective* and *The Market and Infrastructure Perspective* which receives further validation from the technology acceptance model as they focus on individual as well as broader environmental considerations. The economic history perspective is not being vindicated in a significant way as innovations adapt to markets that are not necessarily geographic (Kotler, 2008). This runs counter to *The Market and Infrastructure Perspective* as various geographic locations have nuances in infrastructure. While the development perspective is forward-thinking, it does not however answer why people do or do not adopt but simply promotes equal opportunities. The adoption process has been clearly outlined by Rogers (1994). It involves the stages which individuals go through in making decisions about new objects or ideas. Rogers therefore suggests a linear process which involves knowledge-persuasion-decision-implementation-confirmation. He clearly highlights a process that ends with an innovation being adopted or rejected which has been useful to subsequent research. More recent research has however found that the process of adoption may not be a linear one. While Rogers admits that stages of the process may be rearranged he still maintains that the process is linear.

In contrast Woiceshyn (2000) for example argues that within organizations adoption is a learning process which involves observing, interpreting, integrating and acting. While this may seem to also be linear, the difference lies in the fact that these factors may occur concurrently as opposed to moving completely through one stage before it progresses to another. This work also identifies facilitators of the adoption process as: capability, resources, shared values, effort, motivation as well as factors external to the organization. Straub (2009) supports that the process of adoption is not as clearly defined and linear as previously articulated. He highlights that it is a complex, social and developmental process that is influenced by individual perceptions, emotions and context. This immediately points to the importance of personal factors, which is the thrust of this book, and which has only been implicitly addressed in the innovation diffusion literature. The role of other variables which have been singularly used to address adoption would therefore be to shape, or be shaped by personal factors, which in essence renders them to be mediating variables. They however facilitate adoption to the extent that they are complementary in the process.

The more contemporary work of Zhou (2008) categorizes innovation diffusion research as covering three levels: individual, organizational and national. Much of the same drivers of innovation diffusion such as personal, situational, social, socioeconomic, market and infrastructural factors, can be found at these three levels. Zhou further argues that a fourth level known as the intra-organizational level should be introduced which articulated varying levels of adoption within the organization taking into account how persons respond differently to various stimuli. These are termed as voluntary adopters, forced adopters, resistant non-adopters and dormant non-adopters.

This addresses the "voluntariness" of organizational members based on their response to the variables such as leadership, resource use and strategy, and takes into account the components of the diffusion of innovation (Rogers, 2003) and the Technology Acceptance Model (Davis, 1989). Both of these models are related to the broader *Theory of Reasoned Action* (Fishbein and Ajzens, 1975), which states that an individual's behaviour is a result of his/her attitude towards a specific activity and the accompanying social influences. This sums up the debate very appropriately as an understanding of innovation adoption must have its genesis in an understanding of individual adoption without ignoring social conditions. While the diffusion of innovations theory includes social conditions in its discourse there is a weakness in how it addresses individual behaviour from a leadership perspective. An exploration of the ownership/ leadership factor has been lacking in diffusion and adoption research and is only implicitly dealt with in the field. In addition to the seminal work of Rogers however, the Technology Acceptance Model by Davis (1989) has advanced ideas related to adoption drivers with more specific emphasis on individual adoption. Davis (1989) in his technology acceptance model argues that the degree to which a person feels that the technology will require little or no effort determines perceived ease of use. Perceived usefulness on the other hand is the degree to which a person believes that a particular technology can enhance performance such as time and output efficiency (Lederer et al. 1998). Therefore if there seems to be added value to a process, individuals are more likely to accept technology.

Lederer et. al. (1998) support the notion that perceived usefulness and perceived ease of use were the primary factors which led to an intention to use and ultimately usage behaviour. A key weakness in the TAM is that it only focuses on cognitive processes of the individual. This essentially places this technology decision model within micro decision-making models. The importance of combining the influence of personal factors as well as social systems which was previously articulated makes the TAM very limited in its approach. Perceptions of usefulness and ease of use are often influenced by attitude (Wöber and Gretzel, 2000). They state that usage is influenced by technology attitude. The idea of technology attitude is critical to the discussion and forms an important part of the adoption debate. Theorists focusing on attitude and personal behaviour have however looked at personal usage intentions without sufficiently considering the impact of macro issues. It may be argued that perceived usefulness, perceived ease of use and attitudes are affected by demographic factors such as age and education. Legris et. al. (2003) put forward the argument that in addition to these personal factors, consideration must be given to social influence factors. This research aims to look holistically at micro and macro issues affecting adoption.

Another important personal factor is the role of learning in adoption. Bagozzi, et al. (1992) developed a discussion of the importance of the role of "learning to use" the computer in the overall adoption process. This was an important development in the literature as previous arguments tended to focus in a limited way on the act of using computers. They posit that inadequate learning can affect adoption even where systems are well designed. The discussion of learning in the adoptions process brings into sharp focus the beliefs, attitudes and intentions of individuals. Primarily their work focused on attitude however and addressed attitude toward success, attitude toward failure and attitude toward the process of goal pursuit. This separates the "intention to try" from "actual trying." Learning appears to bridge the gap between intention to use and sustained usage. An individual's intention to use may be signified through the purchase of a computer for example, however without knowing how to use it the computer may simply become useless furniture. The organizational implications are monumental as there must now be a focus on influencing the motivation to learn. This must be coordinated skillfully across varying personal attitudes towards learning and adoption. This brings to the fore the importance of compatibility which is defined by Rogers (1983, p. 15) as "the degree to which an innovation is perceived as being consistent with the existing values, past experiences, and needs of potential adopters."

In contrast to the proponents of the TAM, contemporary researchers such as Dulle and Minishi-Manjanja (2011) found support in their research for the Unified Theory of Acceptance and Use of Technology Model. This model identifies in addition to personal factors such as attitude and expectancy about usefulness and ease of use, that facilitating conditions and social influence are important drivers of usage. Previous theorists (Hooff et al., 2005) have added the element of the impact of situational influences on communication technology usage. They argue that the social environment proved to be influential in generating feelings of perceived applicability. This indicates that the observation of others using or refusing technology in that environment may shape individual views about its applicability to their needs. This discussion is not new, but rather solidifies and extends the early work of Rogers' (1983) diffusion of innovation theory, which recognized social influences as important. These two seminal pieces of research (Rogers' 1962; Davis, 1989) identify contrasting influences and discuss how they affect perceptions for individuals and groups. While the diffusion of innovations theory emphasizes a broader set of influences the TAM focuses solely on cognitive or psychological influences. For the purposes of this book, this makes the diffusion of innovation theory more appropriate given the holistic nature of the study.

Fuchs et al. (2009) further extended the debate and highlight that the environmental context and the availability of ICT infrastructure are important drivers of technology adoption. Lippert and Davis (2006) previously argued that these types of considerations are limited and still do not explain why 50% of implemented systems in the United States for example are considered failures. They therefore introduce trust into a conceptual model and propose that "technology trust" and "interpersonal trust" along with "planned change activities" will influence technology adoption behaviour and enhance the level of adoption and internalization within firms and individuals.

While the approach of these authors introduces some novelty, the focus has not shifted outside of the broader categories of situational and personal influences which were previously discussed by a number of theorists (Rogers, 1983; Davis, 1989; Lederer et al., 1998; Wöber, and Gretzel, 2000; Legris et al. 2003; Vishwanath and Goldhaber, 2003). The above categorizations take into account a multiplicity of factors such as media exposure, compatibility, perceptions, social factors and socioeconomic factors. These have largely been the parameters of the discussions surrounding technology acceptance coupled with discussions regarding mediating factors such as training that mitigate against negative personal and situational influences (Bagozzi et al. 1992; Marler et al. 2006).

Inherent in the weakness of the TAM in addressing situational influences, is the fact that it ignores critical arguments in the resource-based, strategy, ownership and leadership literature. While situational factors may be said to involve these variables, a more detailed exploration of how all of these factors interplay is vital for a more in-depth analysis of why specific categories of firms in particular decide to engage in technology adoption processes. Based on the innovation interdependence concept (Moital et al. 2009), the level of technology adoption is particularly critical to understanding what motivates firms to utilize technology at optimal levels such as e-commerce adoption, which speaks to firms ascribing to engaging in transaction based processes using technologies such as the internet. A discussion of e-commerce and technology adoption therefore follows.

There is a school of thought that emphasizes that while the firm may intend to control the adoption of its e-commerce; this is largely driven and stimulated by customers (see for example Zappala and Gray, 2006). They further highlight that some firms which may claim to engage in electronic commerce may simply have a promotional website rather than one that is an interactive, comprehensive transaction-based interface. They support the views of Daniel et al. (2002) that in fact there appears to be staged adoption. *Developers* are described as firms just beginning to develop online information about products; *communicators* are those firms which use email to contact customers; *web presence* represents those which have websites with online ordering facilities; and *transactors* are those firms which have comprehensive services online from information gathering to payment options and sales follow-up (Daniel et al. 2002).

The e-commerce adoption mentioned in the final stage is scarce for firms which were traditionally offline operators at their inception (Lippert and Davis, 2006). While in many cases resource constraints are posited as the cause, Gray (2006) highlights that ICT adoption may in fact be used to overcome resource limitations in order to grow the business. The reality however is that some small to medium sized firms are unable to find these resources initially. Many small companies in the developing regions such as the Caribbean for example, argue that the returns on such an investment are not substantial enough. Nonetheless, it is possible that this reluctance is however more an issue of attitude than one of resources. Zappala and Sarchieli (2006), purport that the attitude towards ICT adoption and the climate for innovation, are key determinants in the adoption of e-commerce. Very interestingly they contrast the fact that online selling has stagnated although online purchasing is on the rise. This signifies that the rate of e-commerce adoption has declined (Houghton and Winklhofer,

2004). Alternatively Mochrie et al. (2006) argue that some firms simply do not have sufficient capacity to develop both the physical and human capital necessary to develop effective adoption strategies.

Firms may choose to reject some innovations based on the above constraints. Rogers (1962; 1976; 1983) states that rejection of innovations may take place at any stage in the innovation diffusion process mentioned earlier and discontinuance may also occur. He further posits that prior conditions affect the process such as: previous practice, felt needs/problems, innovativeness and norms of the social system. All of the above may be seen at work in modern e-business adoption. In particular in the Caribbean the influence of previous practice, innovativeness and social system is evident. This region typically follows very slowly with innovation, primarily due to lack of resources and resistance to technology based on loyalty to traditional ways of doing things, and systems that reinforce these practices. Despite these broader cultural issues Tanpong et al. (2009) however focus more on the organization types and highlight that, for digitally related sectors, business to customer e-commerce becomes a strategic imperative, while for digitally unrelated sectors, B2C is a strategic choice.

The adoption research after 1991 has identified some key drivers in adoption decisions. These may be grouped as customer pressures (Poon and Joseph, 2001; Bigne et al. 2008; Daniel et al. 2002; Dyerson and Harindranath, 2007), competitive pressures (Bigne et al., 2008; Patricia, 2008; Teo et al., 2009; Wesrthner and Klein, 1999), supplier pressures (Beekhuyzen et al., 2005; Buhalis and Deimezi, 2004; Vrana and Zafiropoulos, 2006), industry challenges (Grandon and Pearson, 2004; Kuan and Chau, 2001; Saffu and Walker, 2008; Teo et al., 2009) and company challenges (Bennett and Lai, 2005; Bigne et al., 2008; Heung, 2003; Law et al., 2004; Stansfield and Grant, 2003; Warden and Tunzelana, 2004). Fundamentally, the pressures from these forces internal and external to the firm are driven by the transformative nature of the internet.

There is consensus by many theorists (Rayport 1995; Choi and Stahl 1997; Bakos 1998; Williamson and Scott 1999; Afuah and Tucci 2001; Applegate 2001; Wirtz 2001; Porter, 2001; Rappa 2002) that the rules of competition for established business have been transformed by the internet in three critical ways:

- Information asymmetry reduction,
- Disintermediation and
- Reintermediation.

Information asymmetry on markets has been reduced by the web (Choi and Stahl 1997; Porter 2001; Wirtz 2001). Collecting information on products such as features and price has become much easier without the need for physical, actual visits to outlets. This development results in less time-intensive and less expensive information searches. For the firm however, the greatest benefit is that the capital requirement to operate in a market is significantly reduced. The Web also minimizes market imperfection and facilitates a larger number of players to compete in cyberspace. The Internet means lower distribution costs, larger market share and higher revenues (Law, 2001). Supplier companies which are always seeking to cut costs are exploring the option of convenient market access via the web. Hatton (2004) further argues that given the millions paid globally for commission, pressures from consumers for lower prices and increasing demands for ROI from investors, many seek to remove these payments. Reduced transaction cost and commission present a strong case for the complete elimination of intermediaries (Buhalis and Licata, 2001). This argument identifies the interplay between the firm and the markets, where firms make strategic decisions that enhance profit and consumer behaviour leans towards greater value for money.

The behavior of customers who are already buying goods and services on-line clearly indicates that companies can use internet technology as a digital channel to deliver services to customers (Ghosh, 1998, Zappala and Gray, 2006). An important debate surrounds whether particular markets are very accessible via this route, a missing element in the literature, and this research will attempt to address issues surrounding determinants of technology adoption such as company factors, personal factors and market factors. The extant literature focusing on disintermediation has consistently concluded what Poon (2001) refers to as a new independence which is being experienced by supplier as well as consumer. The Internet offers an effective means of developing a single and sustainable electronic infrastructure for information gathering and business transactions for both travelers and suppliers (Law et al. 2004). This research aims to investigate why some firms do not adopt when there are apparent benefits. Opposing theorists argue that it is the acknowledgement and utilization of these benefits that will create what is termed as reintermediation.

In contrast to disintermediation proponents, a counter claim has been made that discussions of complete elimination are premature (Palmer and McCole, 1999) and that what is likely to emerge is a new kind of intermediary. The "info-mediary" will take the new role of brokering relationships between consumers and producers in the world of e-commerce. Goldsmith and Litrin (1999), state that the new info-mediary must gather information on the supplier and consumer and be able to merge these interests online. In addition to having an online presence, agents will still act as counselors, therefore they should be able to access and process large amounts of information to narrow down choices and match with consumer preferences (Lang, 2000). They must be able to access information, assess quality and provide expert advice. Users often get so much information that they cannot use it in a meaningful way. It is the job of intermediaries to eliminate these problems by providing a platform for information exchange between buyers and sellers, which aggregates the relevant information and brings the appropriate trading partners together.

The ongoing debate between disintermediation opponents and proponents has largely been unresolved, which has led to a broadening of the technology adoption debate and has facilitated a strengthening of earlier discussed classical theories of innovation diffusion and technology acceptance. Broader concepts for consideration have emerged such as the role of socioeconomic and cultural factors in driving technology access as well as attitude. Of particular significance is the fact that the post-internet discourse has heightened the debate on the global digital divide, which brings into sharp focus issues of information gaps, information-inequality and information poverty across and within societies.

The Digital Divide

The notion of the digital divide addresses the degree to which information technology access provides an advantage and disadvantage to some individuals. A review of the literature on the digital divide addresses the research objective which seeks to to investigate the influence of external firm factors such as the digital divide in technology adoption in owner-managed small firms. The concept has primarily been framed using four distinct and contrasting approaches (Sassi, 2005). He argues that the many different views on the digital divide may be placed in four categories:

- *The Technocratic Approach*, which takes the position that the internet is an important means of everyday life and an essential tool for the new economy. The Public sector should therefore level the access playing field. Most importantly this approach assumes that new technology will overcome social inequality;

- *The Social Structure Approach*, which emphasizes the unevenness of internet use across different social structures. Theorists such as Winston (1986), Sparks (2000) and May (2002) argue that the diffusion of the internet cannot be generalized until societies are able to overcome social inequality;
- *The Information Structure and Exclusion Approach* emphasizes social segregation and the process of marginalization. Lash (1994) and Wacquant (1996) argue that poverty drives the exclusion of some groups from modernization processes such as diffusion and adoption of information technology.
- *The Modernization and Capitalism Approach* argues that capitalism drives modernization processes. Castells (2000) discusses this along the lines of social stratification theory and concludes that what is currently taking place is informational capitalism.

What is particularly critical is that each of these approaches has sought to explore the inequality in ICT diffusion. Within each approach there is consensus that inequality exists, and each approach is only different in their deliberation of causes and solutions. This work fits more closely into *The Information Structure and Exclusion Approach* and *The Modernization and Capitalism Approach* as their major proponents such as Lash (1994) and Castells (2000), respectively, are concerned about how the global economic order influences information inequality in various countries. The question surrounds whether information inequality in the Caribbean affects the level of adoption in owner-managed small travel firms in that region.

Much of the debate on the digital divide spans the period 1986-2005, however fairly little attention has been given to this concept within tourism. More contemporary work on the digital divide by Minghetti and Buhalis (2010, p. 278) articulate that "tourists and destinations within developed countries and between developed and developing countries suffer from a multiplicity of technological divides (motivational, physical, informational, etc.), which lead to different levels of digital exclusion." It is evident that the digital divide issues are also present in the travel and tourism context and it may even be argued that due to the interaction between countries which is necessitated by tourism, the digital divide may be more resonant in this industry. While discussions of the digital divide have been active since the 1970s, the debate became more vibrant as theorists such as Castells (2000) argue that the existing inequality and polarization are outcomes of "informational capitalism" which must be consciously addressed through public policy. Norris (2000) also posits that in the face of world poverty the digital divide is likely to continue in the foreseeable future. This is a particularly striking position since three decades earlier Childers (1975) conducted a literature survey which revealed that up to that point more than 700 documents had already been written which addressed information inequality and information poverty. What has been consistent throughout decades of research is that there is a knowledge gap which continues to widen with unequal access to information technologies.

The digital divide has been conceptualized in three ways. The first was the access divide and then subsequently the learning and content divides became more resonant. The argument put forward by James (2004) serves to point to a closing of the divide based on access. It however does not challenge the position of Rogers (2003) which states that while the access-divide has received most of the attention, it is the learning-divide and content-divide among others which will present a disadvantage for some. This argument provides key insights into issues of matching content to audience needs. Innovators and designers of information technology are likely to create content that suits its own audience in the first instance. Even if this content is later customized for external audiences this ensures a lag as the learning curve for external users is initially higher based on the introduction of content that is not necessarily suited for that environment.

This focus on the content and learning divides reintroduces and brings back into focus the discussions of contemporary theorists (Wilson et al. 2003; Gyamfi, 2005; Willis and Tranter, 2006; Stump et al. 2008) surrounding political, socioeconomic and cultural elements in the global digital divide. What is lacking however is detailed analysis of the role of these factors in bridging the content-divide or learning-divide. The discourse primarily addresses issues of access and public policy *ad nauseum*. While access may be considered to be critical moving forward, it is important to note that access does not guarantee use if individuals are uncomfortable utilizing the technology. The role of public policy makers in poorer developing countries may therefore be that of facilitating training given that content issues are largely outside of their purview. The discussions have however also moved to debates about the impact of new media (Mansell, 2002) and more specifically mobile technologies (Stump et al. 2008), which have been diffused rapidly throughout developing countries and has in many ways lessened the disparity between nations.

There has been some consensus over the last thirty years that information inequality exists however the focus of this inequality has shifted. While earlier research tended to focus on ethical perspectives of inequality, later research looked primarily on political and economic dimensions involved in the digital divide (Yu, 2006). An important development in the literature has been a move from what is deemed fair or unfair to more practical discussions of various actors in the divide. This has provided a more robust debate with theorists such as Feather (1998) noting that information inequality is one of the principal political issues of the information society. This argument receives support from other contemporary theorists (Castells, 2000; Norris, 2000; Golding and Murdock; 2001). Other theorists (Gunkel, 2003; James, 2004) disagree with approaches adopted in addressing the digital divide. Gunkel (2003) for example later argues that previous critiques of the digital divide are limited in that they do not identify difficulties in order to provide solutions but aims to uncover underlying antecedents to its existence.

This argument is only partially sound since an identification of underlying preconditions creates an awareness of areas which need to be addressed in bridging the knowledge gap. Indeed maybe the solution specificity being sought may emerge from future research. The counter-arguments by James (2004) however present a solid critique of previous research by Castells, (2000); Drori and Jang (2003). While much of the earlier research has been concerned with the information inequality within societies, there has been more recent focus on the global digital divide which addresses differences across countries.

The general argument put forward by proponents of the global digital divide (Mosaic Group 1998; Castells 2000; Norris 2000; Rogers 2003; Kirkman et al. 2002; Drori and Jang, 2003) is that economically developed countries control ICT development and access and therefore automatically have a distinct advantage. James (2004) however argues that although individual access can be considered limited in some of these countries, there is still innovative use of technologies in some poorer developing countries. This he highlights, has been brought on in some cases through the foreign direct investment of multi-national corporations. Culture will be explored in addition to the digital divide as another external factor which may influence technology usage and intention to use. Subsequently more internally related factors of leadership and resources will be discussed.

Culture as a construct emerged in the anthropology literature as early as the 1800s. It is generally defined as a shared meaning system (Fischer, 2009). Members of a cultural system tend to share ideas related to values, norms and traditions, which influence social interaction within that group. Culture may be conceptualized at many levels: National, organizational, or consumer market related. In meeting one of the work's objectives which is to investigate external firm factors such as culture in technology adoption in owner-managed small firms, the emphasis will be placed on national and organizational culture, and their influence on the individual.

The role of national culture on the behaviour of individuals as well as firms has received significant examination in the literature. The shared meaning system of a cultural group typically focuses on how that group influences its members through values, opinions, beliefs or behavioural tendencies (Fischer, 2009). Values, opinions and beliefs are deemed to influence behavioural tendencies and any deviation from the norms of that group from a behavioural perspective is usually frowned upon (Toomela, 2003). Individuals are expected to act in accordance with their cultural system. This makes the assumption that people from separate cultural systems will act differently based on a different set of core values and lends itself to discussions of cross-cultural differences.

In contrast Craig (2003) recognizes that while research on cross-cultural differences may identify different behavioural tendencies, the same criteria being tested are the same ones being used to assemble the study participants. For example a study looking at a particular demographic already assumes that there is uniformity in that demographic through the influence of the sampling of that group. More recently Heinich (2010) contends that these cultural differences are not imposed by the researcher but rather systems of meaning drive the research so that differences across systems may be identified. These differences may be seen in individual behaviour or in the behaviour of subcultures within a national culture such as organizations.

Research on cultural differences has been championed by Hofstede (1980; 1991) where he introduces cultural dimensions focusing on international differences in work-related values. The model focuses on differences in power distance, masculinity v femininity, individualism v collectivism and uncertainty avoidance. According to Hofstede (1984) power distance refers to the extent to which powerless members within the firm accept the unequal distribution of power. Individualism as the converse of teamwork emphasizes the need for individuals to take care of themselves and family only. Masculinity addresses systems which value money and success while femininity focuses on nurturing and quality of life. The uncertainty avoidance dimension is concerned with the extent to which individuals are comfortable with risk. Each of these elements is applicable to the study as the firms are assessed in terms of leadership with regard to the interplay between owner-managers and staff (power distance), result-orientation or staff empowerment (masculinity) and risk aversion (uncertainty avoidance). The relationship orientation of the society also addresses the theme of individualism/ collectivism.

This model has been given much attention by scholars from a range of fields. Most notably, management scholars have found Hofstede's discourse on cultural differences to be useful in analyzing organizational behaviour in national cultural contexts and similar models have emerged (see for eg. Schwartz, 1992; Trompenaars, 1993). In fact there may even have been an over-reliance on the model (Bhimani, 1999; Harrison and McKinnon, 1999; Redding, 1994). While these authors have identified that there may have been excessive dependence on the model its use has not been diminished and authors continue to recognize it as a template for understanding cultural differences. Two decades after Hofstede's seminal work there was a resonant critique advanced in research published by McSweeney (2002). He argues that the model's methodology is fundamentally flawed and questions whether culture can systematically cause differences in behaviour between people from different countries. This position is quickly challenged by Williamson (2002) who highlights that Hofstede's research which was grounded in the functionalist paradigm was in fact methodologically strong and that McSweeney's own paradigm needs clarity. A critique of any work that assumes uniformity within national cultures is that there is significant observable plurality within nations that produces differences in behaviours. This argument does not in any way maintain that this precludes an understanding of aspects of human actions and thoughts which may be explained by the national culture within which they operate.

The criticism of Hofsetde's four dimensions as being too simplistic in explaining national differences (McSweeney, 2002) in itself acknowledges that these differences are useful for academic consideration because they at least provide a basis for analysis. Contemporary authors (Gelade et al. 2006) continue to explore the impact of national culture on the organization and identify that there are significant relationships between organizational behaviour and some aspects of national culture. More specifically national differences are highlighted as important in understanding multinational corporations and sales patterns (Murphy, 1999), creativity and innovation (Westwood and Low, 2003), business practices (Tchaicha and Davis, 2005), tourism (Reisinger and Turner, 2002), leadership (Elenkov and Manev, 2005) and corporate strategy and business communication (Ulijn et al. 2000).

The work of Westwood and Low (2003) is of particular interest to this work as it identifies a relationship between national culture and innovation. They conclude that culture impacts on creative and innovation processes. In this discourse about technology adoption and the diffusion of innovations this is a critical consideration. It is surprising that the extant literature on culture and innovations tends to focus heavily on internal organizational culture (Sarros et al. 2008; Jaskyte and Kisieliene, 2006; McLean, 2005; Subramaniam and Ashkanasy, 2001) as opposed to national cultures. While organizational behaviour in general has received significant attention within national cultural contexts, the relationship between organizational innovation adoption and national culture has largely been ignored. Perhaps this relationship is more indirectly addressed through the literature on the global digital divide (Wilson et al. 2003; Gyamfi, 2005; Willis and Tranter, 2006; Stump et al. 2008).

Given that this book attempts to establish the importance of leadership as a complementary driver of technology adoption, an assessment of the influence of national culture on leadership is essential. Elenkov and Manev (2005) make the claim that leadership factors have a major impact on top-management innovation influence. They also conclude that cultural context directly influences leadership and moderates its relationship with organizational innovation. They make this claim on the basis that top-down organizations rely on leadership from top executives such as CEOs who make innovation decisions. This view coincides with the literature on leadership (see for example Hitt et al., 2001; Victorino et al. 2006; Ghobadian and O'Regan, 2006; Falk, 2008) which claims that leaders influence strategic decisions in firms, and makes a solid case that innovation and its diffusion are influenced by the cultural norms, values and practices of a country (Gibson and McDaniel, 2010).

Firm Ownership and Leadership

Organizational Theory researchers have extensively explored the relationship between ownership, decision-making and firm performance. Berle and Means (1932) define the ownership of a firm as those having control and majority beneficial gains from firm activities (shareholdings). Interest in this area can be seen in decades of research (Berle and Means, 1932; Demsetz and Lehn, 1985; Cho, 1998; Cole and Mehran, 1998; Anderson and Reeb, 2003; Welch, 2003). Much of this research has however been focused on ownership structures and their impact on performance while the impact of ownership on the behavior of firms has received less attention (Ghobadian and O'Regan, 2006). It is particularly interesting that so few authors until recently have given attention to ownership and firm behaviour when many firms are a reflection of their owners.

Many divergent views exist in the Organizational Theory literature relating to the impact of firm ownership on firm performance. Authors which have addressed this issue, have sparked the debate and past studies may be categorized as those focusing on entrepreneurship, firm behaviour, strategy and family-owned businesses. For example Porter (1990) posited that ownership of the firm influences the goals of the firm. Daily and Dollinger (1992) also state that ownership influences firm size, strategy and processes. These observations have been quite useful in stimulating the debate as owners in many cases have the ultimate decision on key strategic issues which may determine whether a company succeeds or fails. Entrepreneurial behaviors are influenced by various factors. Education, age, gender, number of family members and dependents, formal experience in business, and management skills have traditionally been regarded as influences on management behavior (Wasilczuk, 2000; Singh et al. 2001; DeMartino et al. 2006). However, a considerable body of research has focused on the role of psychological factors such as motivation, personality, and goal or intention to grow; as explanatory variables in the development of management behavior (Hamlin and Sawyer 2007; Frese et al. 2000).

The entrepreneurship literature on ownership and leadership has brought some focus to family-owned businesses. There has been an ongoing debate about how family-owned businesses operate, and also the issues and challenges for small business owners operating in some industries. Kowalewski et al. (2010) posit that firms with family CEOs outperform those with non-family CEOs due to a greater level of social capital and personal investment in the business. Although this is consistent with the work of Anderson and Reeb (2003), there is a contrasting view which states that family ownership negatively affect performance due to issues of too much personal involvement or incompetence of those appointed to lead (Tsao, 2009). This argument shifts the discussion to one about owner-managers in firms.

Owner-managers are those majority shareholders who engage in the daily operations of their businesses (Shailer, 1994). There are some characteristics which are typically associated with this kind of manager. An owner who manages the firm usually does so for small firms (Gallo and Christensen, 2011) and depends on the firm's survival for family survival. This interconnectedness usually results in less bureaucratic decision-making and greater risk aversion which directly influences managerial style. Managerial style represents the manager's preference in decision-making. Unlike strategy, which focuses on a formalized method of how to reach goals, it concerns a broad and informal pattern of decisions in a discretional situation (Covin and Slevin, 1988). The manager's preference in various situations have been represented by the direction of management related decisions concerned with risk taking, change and innovation, and the aggressiveness of the approach used. In relation to this preference, Carland et al. (1984) classified small business managers into two distinct types; entrepreneurs and general small business managers. With regard to building strategies and behavior preferences, entrepreneurs demonstrate a higher level of self-efficiency, readiness for change, interest in innovation, competitive aggression, and desire for achievement than general managers (Carland et al. 1984; Covin and Slevin, 1988). Alternatively, general small business managers prefer to be decidedly risk-averse, non-innovative, and passive or reactive (Covin and Slevin, 1988). Simply put, entrepreneurs pursue an innovative combination of resources for profit and growth, while general small business managers operate their business as an extension of their individual personality or to further personal goals (Carland et al 1984).

Meanwhile, tactical (operational) behavior is regarded as the profile of management activities that would implement the preferences of a manager while operating a business. Therefore, these tactical behaviors would be affected by the preference of a decision maker at the strategic level (Covin and Slevin 1995). In the case of small businesses, these profiles are revealed for different aspects of management decisions. Collecting information and learning new business management skills, applying new technology or developing new products, quality control of products and service, interaction with customers, the treatment of people in business networks, cooperating with other organizations, employing family labor, long-term and papered financial planning and risk taking; have all been regarded as major elements of practical business management activities (Page and Getz, 1997; Thomas, 1999, 2001; Dahles and Bras, 1999; Morrison, Rimmington and Williams, 1999).

Authors (Tsai and Gu, 2007; Walters et al. 2008) have attempted to address the issue of institutional ownership and firm performance. They have found that ownership of an institution is a significant and positive determinant of firm performance. Therefore the earlier gaps in the literature are indirectly addressed. If in fact ownership is so closely linked to firm performance then the idea of owners influencing or preventing strategy falls within these parameters. George, et al. (2005) extend the discussion to include the influence of owners on firm internationalization. They argue quite convincingly that the ownership of Small and Medium-Sized enterprises (SMEs) influences the proclivity to take risks to increase the scale and scope of internationalization efforts. This is particularly important as strategic management theorists (Porter, 1985; David, 2007) argue forcefully that competitiveness warrants business exploration across borders. Research on firm ownership (Greenwood et al. 2007) has also attempted to move away from the earlier propensity to focus solely on public corporations. These authors have extended the work in organizational theory to account for private corporations and partnerships. An interesting finding is that private corporations and partnerships outperformed public corporations due to the increased monitoring and span of control of the owners in these private corporations and partnerships. This runs counter to the work of Durand and Vargas (2003) which suggests that this difference is due to organizational complexity. What is consistent in organizational theory despite differences in approach based on firm selection and assessment criteria, is that owners influence key variables that affect firm performance generally and even more so when they take on the role of owner-manager (Durand and Vargas, 2003; Greenwood et al, 2003).

This discussion of greater involvement by owners in the management of organizations moves to an exploration of the role of leadership in firm behavior and performance. Kouzes and Posner (1987) define leadership as a reciprocal relationship which clearly separates those who influence from those who are influenced. While this definition provides a basic understanding of leadership, Hitt et al. (2001) describes the leader as a catalyst for strategic change. It is therefore critical to assess dominant leadership categories.

In his seminal work more than three decades ago, Burns (1978) introduced the concepts of transformational and transactional leadership. Since then a number of researchers (Karnes and Chauvin, 1985; Bass, 1999; Bass and Avolio, 2003; Singh and Krishnan, 2007) have attempted to expand on these concepts in particular in the area of their measurement. With leadership being emphasized in this research as the key driver of technology adoption in small owner-managed travel firms, a critical evaluation of the two dominant classifications of leadership is vitally important to this discourse. It must be noted that research which focuses on these leadership classifications in the travel and technology context has been lacking. Broadly speaking, transformational leaders uplift the morale, motivation, and morals of their followers while transactional leaders cater to their followers' immediate self-interests. Leadership that creates change is an important element of transformational leadership and provides a framework for exploring attributes of change leadership. According to Bass (1999) a considerable amount of empirical research has been completed since the seminal work of Burns (1978), supporting the utility of the distinction between the two forms of leadership. He further articulated that changes in the marketplace and workforce over these decades have resulted in the need for leaders to become more transformational and less transactional if they were to remain effective. Leaders were encouraged to empower their followers by developing them into high involvement individuals and teams focused on quality, service, cost-effectiveness, and quantity of output. While this research does not directly address whether leaders should be more transformational or transactional in general, it identifies that new initiatives such as the adoption of innovations like technology are contingent on elements of what have been identified as leadership needs in the mainstream. For example the creation of "high-involvement individuals" and cost-effectiveness are required for the successful introduction, implementation and continued usage of technological tools.

There have been a number of issues surrounding the transformational/ transactional leadership literature. Pawar (2003) argues that while transformational leadership has been identified as an important mechanism for organizational change, there have been concerns about how it has been conceptualized. One major issue relates to the relationship between the two types of leadership. There are essentially three schools of thought on this relationship. The early conceptualization by Burns (1978) viewed both forms of leadership as being at opposite ends of the leadership continuum.

The weaknesses of this view were identified by Bass (1985) in a second conceptualization of the paradigm, where he acknowledged that while transformational and transactional leadership processes were distinct, there may be varying relationships between them. The third school of thought on the relationship between the leadership types, which was posited by Hollander (1993) and states that transformational leadership is merely an extension of transactional leadership will receive significant consideration. The primary reason for this is that there are some fundamental activities in firms which simply require leaders to get followers to conform to procedures, while there are others which require significant transformational skills. The two forms of leadership for the purpose of this work are therefore treated as a hierarchical conceptualization because transactional leadership provides the foundation and necessary preconditions upon which transformation and change are built.

A second issue refers to attempts to separate the concept of transformational leadership from charismatic leadership. In his early conceptualization Burns (1978) viewed charismatic leadership as a type of transformational leadership. Subsequently the research (Bass, 1985; Waldman et al., 1990) has identified that charisma is a central element of transformational leadership. It must be elucidated here that the transformational leader is described as moving the follower beyond immediate self-interests through *idealized influence*, *inspiration*, *intellectual stimulation*, or *individualized consideration* (Bass, 1999). The element which relates to charisma is the idealized influence, and was coined by Bass in light of the perception that charisma was used to frivolously describe popular individuals. This was a very useful separation as "idealized influence" connoted a more meaningful conceptualization that this occurs "when the leader envisions a desirable future, articulates how it can be reached, sets an example to be followed, sets high standards of performance, and shows determination and confidence. Followers want to identify with such leadership." (Bass, 1999, p. 11)

All of the four constructs above will be examined, although the most applicable construct in technology adoption, is *intellectual stimulation* because of the very nature of innovation and adoption studies that focus on creativity and new ideas. This construct is displayed when the leader helps followers to become more innovative and creative, which is operationalized and measured through a leader's openness to followers' ideas. The other three constructs of *idealized influence*, *inspiration* and *individualized consideration* (one to one attention to followers), all address the relationship and kinship that followers (employees) attach to leaders. While these all result in followers aligning themselves to the goals of the leader, the only one which addresses a cognitive and behavioral change in the follower is *intellectual stimulation* through a constant encouragement by the leader to generate new ideas. Hence, an intellectually stimulating leader is one who creates an atmosphere for the generation of new approaches and ideas which may emerge from any level of the firm. In order for employees to accept and understand innovative changes in the organization as well as participate in these changes, the leader must be able to stimulate creative thoughts and behaviours.

This intellectual stimulation is needed to transform ideologies about work and innovative processes in organizations. Much more discussion is needed in the literature on the symbiotic relationship between the leader and the context of the firm; a gap which will be addressed in this research. Firstly, varying organization types may require varying degrees of transformational leadership (Pawar, 2003). In much the same way the technology needs of firms are heterogeneous. This highlights that there is a need for research that considers whether leadership that creates change in any process is being imposed in contexts where this may not be warranted. A second issue as articulated by researchers (Pettigrew, 1987; Pawar and Eastman, 1997) is that organizational context may dictate the extent to which transformational leadership behaviours are supported or opposed. They highlight that values and norms which are reinforced by organizational structure, tend to encourage or discourage transformational behaviours in leaders. The focus of this book on owner-managers of small firms, who formulate and implement policies unilaterally in most cases, makes the last point less resonant. This is so because the leaders are responsible for the organization structure which would enhance or restrict their leadership style. This would suggest that in these organizational contexts the leadership has greater influence over the organizational context than vice versa. The influence of leadership on innovation then becomes pivotal to the debate.

Leaders play an important role in determining which innovations to introduce (Victorino et al. 2006). This is significant as innovations enable a firm to stay ahead of competitors (Porter, 1985; David, 2007). Additionally some innovations such as technological ones may assist in providing more efficient operations. Leaders are also resource handlers and their willingness to provide resources will have an impact on quality and performance (Lewis and McCann, 2004). This fits well into the overall discussions of the Resource-based view of the firm and provides an added dimension which suggests that a firm with a broad resource base is not inherently competitive because it is still subject to decisions and directions of leadership. The combination of resources and the effective use of these resources as directed by the leadership of the organization may determine competitiveness.

Resource-based View of the Firm (RBV)

Resource-based analysis of the firm is a strategic management concept that has existed since the early 1950s (Penrose, 1959) but became more contagious when Wernerfelt (1984) explored the usefulness of assessing firms from the resource side rather than examining them from the product side, which was the dominant approach (Porter, 1980). The resource perspective focuses on tangible and intangible assets that a firm owns. Caves (1980) defines a firm's resources as those assets which are tied semi-permanently to the firm. Wernerfelt (1984), highlights that these may include machinery, capital, brand names, in-house knowledge of technology, employment of skilled personnel, trade contracts and efficient procedures. This marked the beginning of a detailed assessment of the internal capabilities of the firm and its impact on a firm's ability to compete. The heterogeneous nature of firms based on resources has received substantial attention. In addition there have been discussions about the competitive edge that these differences provide. Barney (1991) further articulates that these differences may be long lasting and may provide a sustained advantage for some firms. In fact the literature on the resource-based view, whether assessed relative to neo-classical economics (Ricardo, 1817), evolutionary economics (Nelson and Winter, 1982) or to structure-conduct-performance (SCP) based theories of competitive advantage (Porter, 1980), holds that the heterogeneity in resources leads to differences in performance. For example resources and capabilities are referred to by neo-classical economists as factors of production and it is felt that these enable companies to supply more. They purport that this supply is elastic which suggests that once demand increases the price and supply of the resource will increase. Due to the evolution of resources, some resources such as human resources which fall outside of the ambit of these factors of production may be inelastic due to the length of time which it may take to develop these resources (Barney, 2001). This inelasticity of supply may however be a source of sustained competitive advantage since other firms may take a long while to develop these resources (Peteraf, 1993).

Aligning the resource-based view to Nelson and Winter's (1982) work on evolutionary economics would suggest that firms' processes of conducting business vary. Therefore ineffective and inefficient operations as measured by input and output differentials, will decrease a firm's competiveness. The implication is that those with the necessary resources such as financial, human and technical resources to aid their operations are likely to have an advantage. So the varying performances that a routine yields will determine how well a firm competes. The most popular set of discussions surrounding resource-based perspectives however are more closely positioned relative to SCP based models of competitive advantage (Barney, 2001), which more directly advances the notion that firm-specific attributes and resources determine performance and sustained advantage in an industry.

Much of the work of Resource-based theorists (Penrose, 1959; Chandler, 1977; Teece, 1982; Wernerfelt, 1984; Peteraf, 1993; Mowery et al. 1998) indicates that those firms holding resources may maintain a relative position of competitiveness to other firms without those resources, once they act rationally. This idea of acting rationally points to the fact that if there are already holders of these resources this may affect costs and revenues of those who acquire these resources later on. The resource position barrier is only partially similar to Porter's (1995) entry barriers given that it focuses more on barrier situations between those already in the market (incumbents), rather than simply focusing on the ability of incumbents to provide barriers to potential entrants. The resource position barrier therefore provides a basis upon which firms already competing may outdo each other (Kraajenbrink et al. 2010). Firms may therefore use their resource position to cement their lead. What is evident in these discussions is that resources play a significant role in a firm's decisions as it many times dictates what can or cannot be done.

Peteraf (1993) more closely aligns resources with firm performance. She elucidates that the resource-based model concerns itself with the internal accumulation of assets and asset specificity. This of course presupposes that firms are heterogeneous in terms of internal capabilities and resources, which has always been fundamental in strategic management discourse as it has always been argued (Andrews, 1971) that a firm's strategies should emerge from an assessment of its organizational competencies and resources. On the basis of the heterogeneity of the firm it is therefore clear that strategies must differ between firms and must be dependent on what the firm is able and willing to do; a direct result of their asset base. Firms with broad resource bases are therefore more likely to pursue diversification (Montgomery and Hariharan, 1991). This diversification may refer to diversification of products, markets or processes. It enables firms to make decisions for example about what platforms they may or may not capitalize on. This notion of the resource base being directly linked to diversification is also endorsed by a number of theorists (see for example, Teece, 1982; Wernerfelt, 1984; Williamson, 1985; Mahoney and Pandian, 1992). While a firm's resources certainly are not the sole determinant, they allow for a firm to have many more choices between strategy options as their internal capabilities allow for multiple avenues for conducting business.

While firm-specific capabilities have gained remarkable attention, the resource-based view has been extended to look at more than just how firms compete but also how they collaborate. Management Scholars (Teece, 1982; Mowery et al. 1998) have applied the resource-based view to inter-firm collaboration. More specifically there has been an exploration of how the resources of a firm may influence business partner choice. Mowery et. al. (1998) focus on technological capabilities as the concept of assets and firm-specific capabilities evolved with time. They posit that technological overlap plays a role in partner selection. The broader application however is that firms with significant resource bases are particularly attractive for collaboration. The literature does not point however to which category of resources are most attractive and also does not give much consideration to variances between industries.

Fundamentally, resource-based discussions are grounded in debates about a firm's distinctive competencies, heterogeneous capabilities and diversification strategy (Mahoney and Pandian, 1982). They further argue that distinctive competence is a function of the asset base which a firm possesses at any point in time. The ability therefore to prolong or sustain this distinctiveness will depend on the type of resources and the type of industry. The RBV is not intended to provide managerial prescriptions (Barney, 2001). While this is so, it points very clearly to implications of why some firms have a strategic competitive advantage over others and can be of great value to managers (Kraajenbrink et al. 2010). Brouthers et al. (2008) have extended discussions of the resource based advantages to the international context. They suggest that differences in the institutional environments of nations may influence its applicability. Consequently the resource based perspective may become more robust when the moderating influence of national institutional environment is also considered. Context specificity should play a role in assessing firm performance. This does not preclude the application of the RBV but rather suggests that firm-specific capabilities and resources be considered within broader institutional contexts and firm strategy considerations.

Emergent Conceptual Frameworks

The specific gaps have been highlighted in the previous sub-sections however the general gap in terms of approach, points to the exploration of simplistic relationships between a limited set of variables. At present, very little is known about the complex relationship between ownership/leadership and technology adoption. In fact, very little research has gone beyond singular relationships between these personal factors and firm strategy and resources. The depth of the discussion has therefore been insubstantial as it neglects the complex interconnectedness of a multiplicity of variables.

The key argument of this research is that firm technology adoption is affected by multiple factors working concurrently and these are highlighted through a distilled framework which highlights a number of antecedents such as: Ownership/Leadership, Firm Strategy, Resources, Culture and the Digital divide. This framework is later framed within the theory of Organizational Decision-Making to identify the key driver as being ownership/ leadership. This work therefore explores how these personal factors are influenced by macro issues such as culture and the global digital divide, and in turn influence firm strategy and resources which ultimately affect technology adoption decisions in firms.

Both iterations of the conceptual frameworks which have emerged from an analysis of the gaps in the literature may be seen below in *Fig 1-1 and Fig 1-2.*

Fig. 1-1 Firm Technology Adoption Framework: First Order Iteration

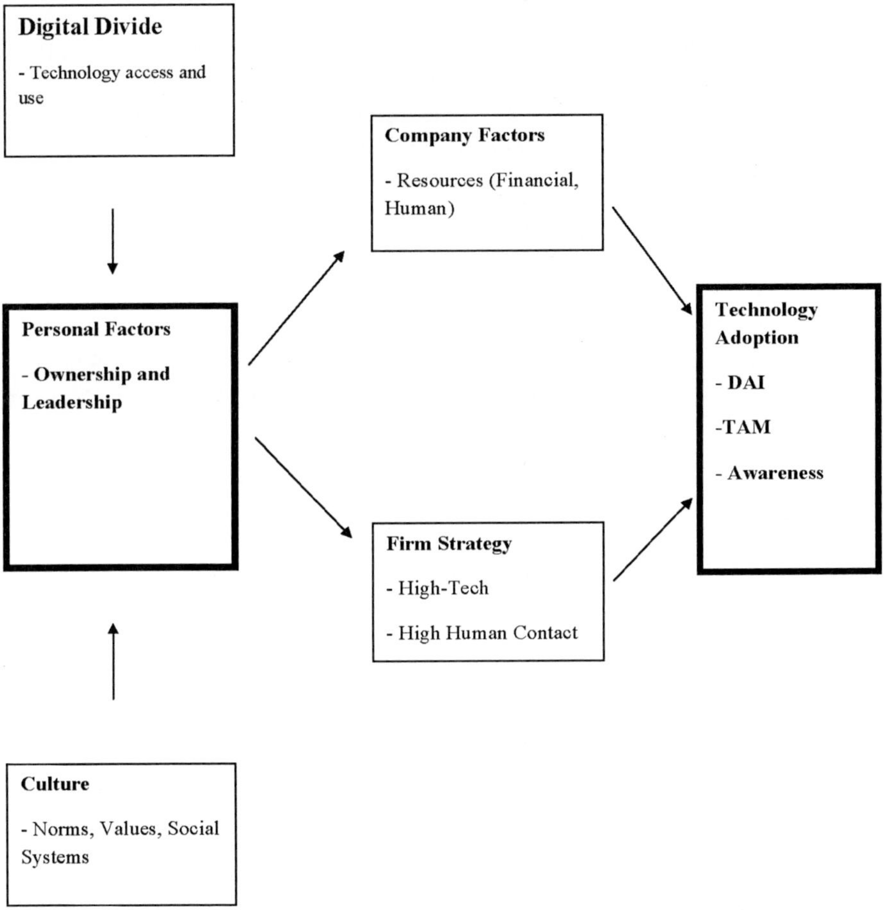

Fig 1-1 identifies fragmented relationships which exist in the literature which have typically focused on singular relationships. The researcher has however brought these together to demonstrate that where one factor directly influences another factor it ultimately indirectly affects other factors in the technology adoption discourse.

There have been some gaps in addressing technology adoption in firms due to this fragmentation in previous works. *Fig 1-2* illustrates a coherent framework, which places technology adoption decisions within the broader theory of decision-making and highlights relationships which may be explored for future research. This research further emphasizes the leadership role as the focus is on owner-managers of small travel firms, who have a great deal of autonomy in decision-making. This second order iteration is created as an input framework to be tested based on the literature and the gaps in the first order iteration.

The conceptual framework above in *Fig 1-2* weaves together the various concepts discussed in the literature review by establishing that the path to technology adoption within firms involves multiple factors. Factors which are external to the firm such as culture and the digital divide provide context parameters within which firms must operate. This influences how the ownership of firms evolves and even more so how leadership attitudes are formed through primary socialization as well as through education. The types of entrepreneurs and leaders which emerge will directly influence how firms are structured and the strategic choices which are made and supported by resource allocation. These factors drive the behavior of the firm, which includes innovative behaviours. This is a product of organizational culture, which is developed through strategy and policy decisions. The research on firm behavior identifies a clear relationship between how a firm behaves and ultimately its performance as measured by its profitability and competitiveness. The literature points to systematic approaches to assessment of firm performance which are geared towards informing strategy formulation, which will either support previous technology decisions or result in changes to previous decisions, as demonstrated in the feedback loop of the framework.

Fig. 1-2 Technology Decision-Making Input Framework

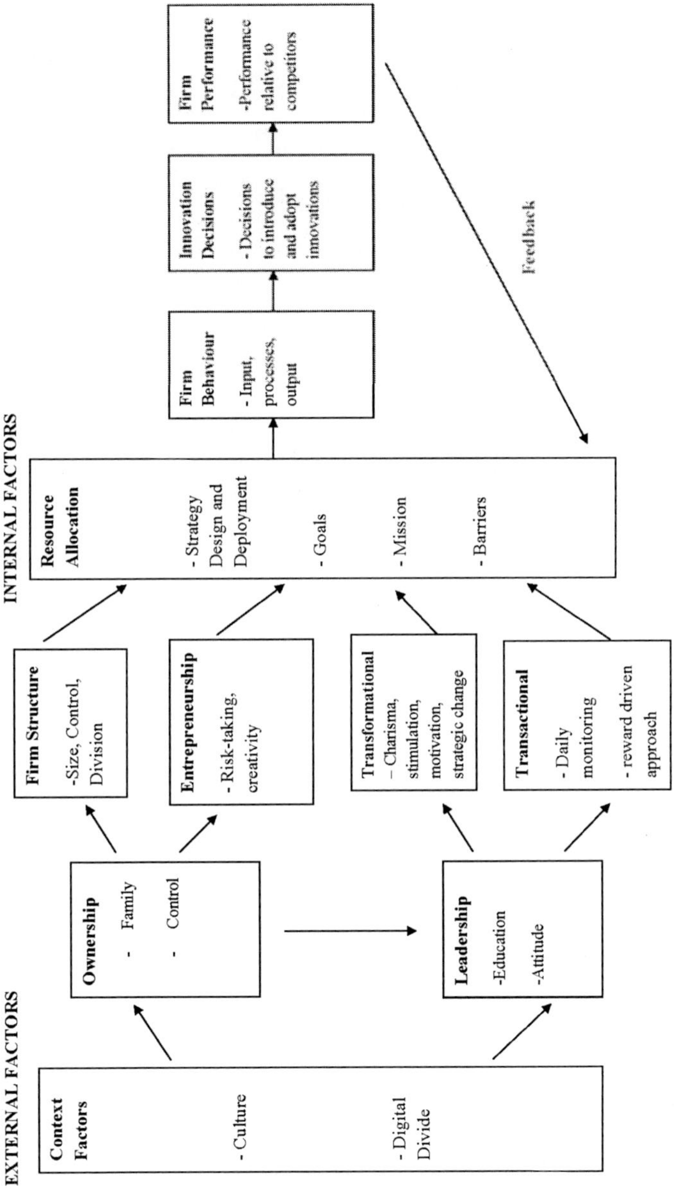

CHAPTER TWO

INTERNAL FIRM FACTORS: A RESOURCE-BASED VIEW OF JAMAICAN TRAVEL FIRMS

There are 43 travel agencies in the island of Jamaica. Thirty one (31) of these firms were interviewed in phase 1 as they shared the same characteristic of having owner-managers (those owners who engage in the daily operational activities of the firm). Owner-manager firms were targeted as this reflected more autonomous decision-making in the firms. These firms varied in size and length of operation. The general findings from the exploratory closed-ended questions indicate that the firms in the sample typify older firms as 21 (67.7%) of those firms were in existence for more than 15 years and only 5 (16.1%) of the firms started operations less than 10 years ago. Using the number of employees in the firm as a proxy for firm size, it was discovered that 87.1% of the firms had fewer than 15 employees and 71% had less than 10 employees. The firms may therefore be categorized as mature, small firms.

The age of these firms may point to a possible cause for their resistance to new innovations. Relationships have been identified between innovative activity and the age of the firm (Lynskey, 2004). One side of the debate identifies that age and experience may lead to more competitive behaviours while the opposing argument points to newer agile firms being more open to cutting edge approaches. The latter would seem to be more applicable in this research as owner-managers in older firms highlighted that their traditional methods continue to work therefore there is the perception that the need for change is not evident. One respondent stated: "We have been in this business for almost 20 years and we have survived harder times than these. For the last 10 years we have used SABRE and it has worked well for us. I think that these challenges will soon pass just like they did when we faced commission cuts in 2001."

In terms of the markets being served by these companies, it was striking that all of the firms in the sample served the outbound leisure and business markets but none of the firms served inbound leisure or business markets. Reasons given were that inbound travel to the island does not pay a commission. The excessive emphasis on outbound travel highlights the workings of a relationship oriented society. The client bases of these firms have been fairly constant with little new business being generated. Respondents stated that they served most of their clients for over 10 years and that they felt that there was a high level of loyalty. In the same vein the respondents recognized that the client base was decreasing rather than increasing, as their clientele (primarily baby boomers), seemed to be booking less through them while Generation X and Y were seeking alternative means of booking online. One respondent stated: "Most of my customers have been with me for a long time and now they seem to be booking less, which is maybe because of the global economic crisis or just the fact that they are older than they were and are naturally traveling less. At the same time we tend not to get a lot of young people booking with us because they probably do it themselves."

In order to reach these markets which they serve, 100% of the agencies mainly rely on Global Distribution Systems (GDS), primarily SABRE and AMADEUS, who provide them with the systems. On this basis 96.8% of the firms considered their technology use to be very high. While the majority also felt that technology is important for sales and marketing, they were unable to see the importance of any ICT outside of the GDSs. As a result 77.4% of the firms felt that company websites were unimportant and an even larger percent (87%) considered the online market to be very unimportant. As a result none of the firms had a Facebook, MySpace or twitter account, and the internet is primarily used for emailing.

Of particular interest however was the fact that all of these firms agreed that the variables identified by Eason (1998) were applicable. Over 90% of all respondents agreed in all instances that the internet promotes faster work, makes work simpler, takes less effort and improves output quality. However, firms were overwhelmingly uncertain or in disagreement that the internet improves their image among competitors and saves money for the company. Therefore, while they could identify operational benefits of the internet, they were unconvinced that it was helpful from a strategy and resource perspective. To further highlight this point, all the respondents felt that an internet investment was either medium (45.2%) or high risk (54.8%). This suggests that there was some ambivalence or at least uncertainty about whether returns on such an investment were worth it. This refers to three types of perceived risk: psychological risk (Eastlick and Lotz, 1999) time risk (Tan, 1999) and financial risk. All three were applicable as owner-managers were apparently worried about the risk of losing time and money but also about the psychological risk of becoming dependent on the internet. As stated by one respondent: "I have seen too many companies in other industries become so technological that they lose some of their customer service quality and I do not want my company to become so dependent that we forget who we are."

Most importantly firms highlighted that their strategic decisions were driven by customers and the variables used by Wöber and Gretzel (2000) such as speed, convenience, accuracy and security were all important to the majority (80.6%) of their customers. The only variable which 100% of the respondents agreed was critically important to their customers was personal interaction, a service which they articulated, could not be provided by the internet. Further exploration revealed that all of the agencies use computers for back office accounting and administrative functions through software called "pear tree". This is an automated system which allows for payables and receivables to be handled efficiently. For front office operations 100% of the agencies use Global Distribution Systems (GDS) to serve their markets. This is very significant in highlighting their attitude to investing in technology. One respondent stated that: "The airlines give us these systems for free as well as training in how to use them so we do not have to invest too much or face the risk of losing money when things do not work out"

The most popular GDS among firms was SABRE due to the initial relationships with American Airlines who were the owners of this system before it became independent. AMADEUS is next in terms of GDS use and is growing in popularity among the firms. Very few firms saw the need for any other technological tool in their sales and marketing effort. They also pointed out that their competitors who created websites had not gained any benefits from doing so. As one respondent highlighted: "I have seen some travel agencies spend a lot of money to design a website and I cannot see how their company is better off than mine right now. It just seems like a waste of money."

For those who created websites in the past they highlighted that a website seemed like a great idea because it would bring visibility to the firm. Having created it however they recognized that they did not have the personnel to update and maintain such a website. One owner-manager pointed out that "Creating a website is one thing but maintaining it is completely different matter". These websites presented information in the initial stage of the operation, much of which is now outdated. Of the firms which had websites, only two of them had the intention to engage in online selling. Some were interested in visibility and providing general information about the company and where it is located while others went a step further to provide marketing information such as package deals. In most cases however, customers had to make the booking on the telephone or at the office. This would seem to suggest that the intent was to use the website to influence traditional offline bookings. One respondent pointed out that: "When designing the website we did not consider doing complete bookings online because that is complicated and risky. We just wanted to become more visible so that potential customers would see us and call us to make bookings."

This approach was taken because of the additional cost of familiarizing employees with the technology. The issue seems to be one of learning to use the internet (Rogers, 2003; James, 2004; Stump et al., 2008). Issues of ease of use (Davis, 1989) may be addressed by greater internet training. Notably these owner-managers overwhelmingly saw little importance in implementing this kind of training. This was because they preferred to use systems where training was already provided free of cost. As stated by one respondent: "There would be additional cost in training employees and I am not sure that we would increase revenue as a result. At least with the GDS training is provided at no additional cost to us so even if there are no significant returns then we are not seriously affected."

This is evidence of a greater operational focus than a strategic one by these owner-managers. They are unable to see the long-term benefits of increased training in the use of internet technologies and choose to allow for technology training only if this is free. Image among competitors and monetary savings were used to address issues of strategy and resources which are factors from the literature, which influence the adoption of innovations. Image was the only item in this section on relative advantage for which there was no uncertainty. There was a high level of disagreement as 80.6% of the sample disagreed or totally disagreed that such a benefit existed. This was due to the fact that they only viewed local operators as their competitors and these companies were equally as ambivalent about internet use especially for front office functions. One respondent pointed out: "I do not think that our local competitors are using these technologies either and so whether I use it or not they will not think I have a better image. In any case they do not know what I am using or not using and I don't pay a lot of attention to what they use."

While the competitors are also not using internet technology for sales and marketing, owner-managers are being narrow in their approaches. In actuality, the fact that competitors are not using this tool for sales and marketing may present a strong case for why its use would probably yield a competitive advantage. The most telling of these benefit items is monetary savings. It must be reiterated that these are small firms with limited resources. The management structure for each of the firms is fairly simple, where the owner is the manager and provides supervision for a few employees. The returns on investment therefore feed directly into the owner-manager's livelihood. One owner-manager stated: "My company is my life investment and so anything that I spend on for the agency must guarantee that I either make more money or save some money. I am really not sure that I would save or make any money from increased internet use." Concerns regarding competitive image and efficient use of resources are directly related to the strategy and resource allocation debate, which will emerge subsequent to the case study analysis, which encapsulates the characteristics of the highest and one of the lowest adopters.

Case Studies

Case A

The firm and owner-manager in Case A represent a minority of high-adopting firms. The internet activity in this firm involves regular emailing, online browsing and minimal online sales with an intention to broaden this platform. The owner-manager of this firm is an avid user of the internet for personal online buying and personal social media use and has three teenage children who are avid users. He cites this as being influential in his decision to introduce an online platform in the firm. He states:

> "I buy things online all the time like gadgets and books and I have always felt that it is an excellent way to reach customers. I have delayed using it in the business for a while because I figured that my client base was not ready. I think the time is right at this point as the client base is becoming more savvy. In any case I can try to reach people outside the current client pool."

> "I currently use Facebook to connect with friends and I am not using it in the business as yet because I need to find out more about whether customers view it as a serious business tool or just as something to do for fun. I am open to the idea though and will look into it."

The experience with using technology has impacted this manager's proclivity to introduce some forms of cutting edge technology in the business. It creates a level of openness to exploring options. This openness was further influenced by a university education in business studies, which allowed for an understanding of various business models.

> "Back in my days at the University of the West Indies I learned how to analyze market needs on a broader scale through needs assessments. This kind of business knowledge can be applied to any industry and is not limited to just travel."

> "I also had the privilege to work in banking and finance as well as real estate, and both of these industries faced hard times so I understand that in difficult times, which as you know the travel agencies are facing, we need to find new strategies to survive."

Previous work experience in other industries also played a role in the outlook and approach of this manager and he stated that this made him less risk-averse as he noted that "even though I live in a society that is not very open to technological investments, something new must be tried in times of crisis." The most profound statement made by this leader was that "you can't expect to do the same thing over and over and expect different results." This owner-manager may be characterized as a market leader who aims to take a proactive approach rather than a reactive one and seeks for new opportunities to counteract imminent challenges.

Case B

This case represents low adopting firms and echoes the sentiments of the firms in this category. The level of internet use is limited to emailing and the main tool for sales is the GDS which is given free to these firms. The interest in spending money and making an investment in technology and the internet is therefore very low. The owner-manager has been in the travel industry for the last 25 years and believes that the technology revolution has affected service quality.

> "When I first started in this industry I used to work for Air Jamaica and then I went to Martin's Travel. Afterwards I opened my own agency because I was impressed with the level of service of these companies. Back then we didn't have all this online craze and we believed in personalized service."

This manager feels that there is a great deal of incompatibility with technology and having a personal touch in business transactions and believes that if she continues to offer exceptional service then her business will rebound and once again become profitable. Her view is narrowly focused as she entered the travel industry straight from high school and has not had any formal educational training since then. She is therefore limited by her experience and refuses to do anything differently especially without any evidence of the benefits of significant technology use. She has one child who is now an adult and does not live at home. She is risk-averse to the internet and highlights that:f

> "I would be wasting my time to try to sell online to my clients because most of them do not like using the internet to do serious business; furthermore they like having that personal contact and they wouldn't trade that for anything."

The personal characteristics of these owner-managers at opposite ends of the spectrum play a role in driving strategy and resource allocation decisions, which are the key internal elements of the firm in technology adoption studies as previously highlighted in *Fig 1-2*. Each of these factors will be discussed based on the findings from the qualitative research.

Firm Strategy

Overwhelmingly strategic management theorists have relied on the work of Harvard scholar Michael Porter (1985) in discussing business strategy in competitive environments (see for example David, 2007; Kim et al. 2004; Stonehouse and Snowdon, 2007; Poon, 1993; Wynne et al. 2001). A resonant argument in this body of research is that in formulating strategy the firm must be guided by competitors as well as consumers. It is expected that the result will be a well-informed, robust and dynamic competitive strategy. While the bottom line of strategy surrounds competitiveness, strategy will be conceptualized in three ways for the purposes of this research to reflect the supply chain. This will include overall competitive strategy, procurement strategy and distribution strategy. Most importantly each of these will be examined for their role in technology adoption in these firms.

The findings largely support the work of Porter in terms of strategy. According to Porter (1985; 2001) a company develops its business strategies to obtain competitive advantage over its competitors. It does this by responding to five primary forces:

- The threat of new entrants
- Rivalry among existing firms within an industry
- Threat of substitute products/services
- Power of suppliers
- Power of buyers

Porter suggests that technological change, such as ICT, is one of the most important forces that can alter the rule of competition. This is because a majority of activities in an organization create and use information and this will be more powerful for industries where information is the key product. Until recently travel agents have been the key intermediary with information as their primary stock in trade (Cheyne et al. 2006). They argue that the Internet is a major source of information and so the agencies' stock is at risk due to the availability of information to end users in an unprecedented manner. This brings into sharp focus a need to understand the role of technology in restructuring the industry and even more so the drivers of adoption for particular types of firms in this process of revolution.

In relation to the five forces previously mentioned, the firms in the sample were not particularly affected by the threat of new entrants since there continued to be a reduction in the number of firms operating in the environment. Even with new entrants such as Expedia, the firms did not view firms outside of the local marketplace as competitors. Rivalry among existing firms was however a very important element. The major strategy for competing with rival firms seemed to be the undercutting of service fees which led to high levels of mistrust. One respondent stated: "As a group we decided that service fees should be J$2,100 and there are some companies who have secretly been charging as little as J$1,500. At this point we cannot even trust the association to regulate the industry." The Jamaica Association of Travel Agencies (JATA) includes all of the firms in the study. In response to the commission reduction by suppliers, the collective decision was taken to set service charge at a minimum level. In order to increase market share some firms have chosen to informally sell services below the price floor to which they had agreed. This short term approach resulted in rival firms doing the same and this led to a lack of trust and the absence of long term strategies to gain a competitive edge. An interesting response from an owner-manager revealed that: "The president of our travel association (JATA) is the manager of the largest travel agency which has been buying all of our small firms. To have someone like this as the president means that the association is no longer looking out for the best interest of smaller, under-privileged firms like mine"

The greatest threat of substitute services emerged with the advent of the internet and in particular its use for purchasing travel services. A striking finding was that the perception existed in these firms that their services were not completely substitutable. This was based on the view that they provided a personalized service, which could not be provided by the internet. A common sentiment was expressed by one respondent who said: "my clients will never get this level of one to one service online". Another stated that "they always come running back when they are disappointed with online transactions". Those firms in the minority which were exploring online options for sales and marketing held the view that markets were beginning to care less about this personalization and decided that a multiplatform approach was the best strategy.

The power of suppliers and buyers is very evident in the study. The major suppliers are airlines and their actions resulted in an alarming decrease in the number of agencies with the number moving from 105 to 43 between 1999 and 2009. In 2000 American Airlines, British Airways and the regional carrier Air Jamaica cut commission from 9% to 6% for travel agencies in the region. As was previously noted, when Jamaican travel agencies attempted to boycott the sale of American Airlines tickets in response to the 3% cut in commissions in 2000, their efforts proved futile as the airline, which accounts for approximately 70% of air traffic to the region, simply decided that direct bookings was the best course of action. The agencies' position was further weakened when Air Jamaica and British Airways followed suit with similar cuts just two months later. Further commission cuts were experienced in 2009 from 6% to 3%. The power of buyers has also been seen in the overall decision to remain in operation with low levels of technological innovation in response to the perception that the buyers are not particularly interested in high technology platforms. This necessitates an exploration of the distribution strategy in these firms.

A resonant theme from respondents surrounded whether some forms of technology adoption were compatible with their sales and distribution strategy. This they articulated is driven by consumer needs and preferences. Of all the variables posited by Wöber and Gretzel (2000) such as speed, convenience, accuracy and security, the entire sample recognized that the markets which they served valued personal interaction as the most important variable in conducting travel business. Respondents spoke in particular about high-end clients who had 24/7 access to them in much the same way that they would have access to online booking sites. Relationship building forms the cornerstone of their operation. One manager stated:

> "My clients can call me anytime of the night in the same way that they can turn on their computer. I have had situations where a certain prominent businessman called me from China to change his travel plans because he had an unplanned meeting. He was not interested in changing anything himself so relationship is important to these high-end customers."

While respondents agreed that the internet provided speed and convenience they were less convinced about its strength in terms of personal interaction, accuracy, security and simplicity. They felt that these were all areas in which traditional agents had a competitive edge over online sources. Typically they felt that their local outbound market found online transactions to be difficult as a result of trying to do so and making mistakes, which influenced their view of its accuracy. Additionally they felt that the market did not think there was sufficient credit card security. The interviews revealed that:

> "Our customers are not yet ready for online buying for a number of reasons. First they are not familiar with the technology, then many of them do not have credit cards and those who do, don't think it is safe to use it with someone that they cannot see."

> "Too many of my clients try and make mistakes and then want me to fix the problem which they have created. Sometimes I charge them to do this so that the next time they will think twice about trying to do something that they are not competent at."

> "The customers that I serve do not want too many options. This is the problem with the internet. They see too many options and get confused. This is why I think they need me to make it easier for them."

One element which emerged from the interviews, which was a variable that was not found in the literature was accountability. Owner-managers stated that their customers liked to have someone to blame if things went wrong. One respondent stated that "they like to have someone to quarrel with if their booking is messed up and they can't do that with the net." If there was someone accountable for errors then that person would be responsible to correct those errors. In online situations they felt that their "low-tech market" experienced what has been termed as transaction loneliness.

Of particular interest was the fact that most managers felt that there is a high level of incompatibility with having a high-tech strategy coupled with a hi-touch one (intense personal interaction) despite compelling arguments by theorists (see for example Buhalis and Licata, 2002; Hatton, 2004) that a multi-platform approach to distribution will yield sound business models. The owner-managers in the sample seemed more concerned with what Eastlick and Lotz, (1999) refer to as a perceived psychological risk of becoming too dependent on the internet and losing the essence of their customer interaction. One owner-manager stated "if we rely too much on the technology our business loses its personality." It is clear therefore that if these owner-managers are to engage in greater internet technology adoption, they will need some form of assurance that online and offline channels of distribution can "peacefully co-exist". In determining what channels to use to distribute services to customers these firms were operating on limited information. Firstly none of the firms in the sample knew what distribution methods their competitors were using and so their claim that customers could not handle some platforms was unfounded. Secondly they focused primarily on current platform uses of the client bases and typically ignored the changing needs of the existing customers as well as the new needs of the emerging markets.

The firms under examination are intermediaries and therefore do not operate at either end of the supply chain. It is important to look at more than their strategies for distribution and analyze their approaches to buying from suppliers. The concept of strategic purchasing is not a new one and theorists (Eames and Norkus, 1988) have long discussed relationships in business to business (B2B) transactions. They have in fact suggested that business buyers should create a strategy that consists of four elements: self-analysis, vendor-analysis, vendor-performance analysis and programme review. A more contemporary discussion of B2B relationships has been carried out by Ryals and Humphries (2007), who state that while unnoticed to business marketers, there has been a quiet revolution in supply chain management where the traditional emphasis on least-cost transactions has given way to a focus on long-term relationships with a few key suppliers. This again is in contrast to the previous views of authors (Ricketts, 1994; Williamson, 1979; 1990; Reve, 1990) and demonstrates a shift from the view that transaction cost is the key driver.

In the case of the owner-managed small firms in the sample, there is no clear procurement strategy for two major reasons. The first is that the pool of suppliers is fairly small for the type of air travel being demanded (ie. from Jamaica to the United States). Secondly there is very limited supplier diversification into areas such as cruises, hotels and tours. Cruises and hotels in particular are now offering five times more commission than airlines in the Caribbean. This diversification into accommodations and cruises would also allow for firms to re-invent their level of importance through a greater influence over inbound travel. It may be categorically stated that while procurement itself has an effect on these firms who use GDS systems provided by the suppliers, the current situation does not encourage the development of a procurement strategy that would influence innovative behaviour. These relationships for the most part have remained static in terms of the processes for the procurement of these travel services. The firm strategy debate in the context of technology adoption for small owner-managed firms must therefore be assessed primarily in the areas of competitive and distribution strategies. These have however emerged as the effects of broader factors such as leadership in firms with these characteristics.

Leadership and Strategy Formulation

There are some unique characteristics for the firms in the sample. They are small, owner-managed firms in an information intensive industry. All of these characteristics have influenced which factor has emerged as the most dominant one in this research. Owner-managed firms typically provide a great level of autonomy for the owner in the daily operations. They tend to act not only in the best strategic interest of the firm but also for personal and family interest.

The small size of these firms usually signifies inherently simple organizational structures in terms of reporting relationships. Once again this leads to a greater autonomous capacity for the owner-manager who in many cases for this sample did not have a board of directors to whom they were accountable. Decision-making was typically unilateral and often reflected the personal preferences of these leaders. This is particularly critical in an information intensive travel industry where firms are using relatively low levels of cutting edge ICT. When asked how decisions about what new innovation to introduce were made the responses included:

> "If I see the benefit of some new innovation then I will check to see if I have the money to introduce it. If I am convinced that it is something that I absolutely need then I slowly introduce by telling employees about it and then doing orientation about it."

> "Well I make the decisions since I am the one who will be answerable to my family if the business fails. I sometimes ask for the advice of friends who may have some experience with a tool or with a similar business but at the end of the day it comes down to my decision."

Strategy formulation in these firms is usually within the sole purview of the owner-manager and these strategies tend to reflect the experiences of the individual. As indicated in chapter 6 on leadership, the decision to adopt a high-tech or low-tech strategy was directly linked to the level of personal technology involvement of that manager. In the cases where leaders were not experienced in the use of a particular technology those firms tended to use more traditional approaches. Firm leaders who were simply using the GDS as the highest form of technology in firms responded alike in terms of their personal technology use. Some responses were:

> "I have to make decisions that will enhance my firm and I am not comfortable introducing things that I am not familiar with. If I use the technology and I am comfortable with how it works then I might use it for the business. I don't use most of these new technologies myself so I am not sure if they are right for the company."

> "Most times I don't use these technologies because I am not good at using them. I have survived all of my life without using some of these things. The company has also been around for more than 20 years and it has lasted this long without these new gadgets."

Where firms had technologically savvy leaders there was a greater propensity to create distribution strategies which included certain technologies. The most technologically inclined firm had already developed an online selling platform and was rejuvenating its use while researching and exploring social media as a marketing tool. The manager of this firm stated:

> "The power of the internet cannot be ignored. Markets are changing. As a consumer I use the internet to purchase books, music and travel products and so I know that in the same way that this works for companies that I buy from, it can work for my business too."

There were other important variables such as education, family status, intellectual stimulation which determined some leadership typologies which are discussed in chapter 6. The purpose here is to establish that leadership was the dominant factor and that strategy was not a central driver in the technology adoption decisions of small, owner-managed firms.

The strategic choices emerged as more of an effect rather than a cause. They were the result of certain types of leaders. Leaders who made strategic decisions which involved the adoption of new innovations were typically those leaders who employed agile strategies. On the other hand those leaders who were inflexible were naturally more resistant to strategy formulation which included substantive or structural changes. It was generally found that the owner-managers were strategically weak as there was little vision or long-term planning for the firms. This may explain the lack of interest in transformational tools such as the internet which would require considerable deviation from the norm.

In addition to strategy, resources have been identified as an internal influence on decision-making processes in organizations (Kraajenbrink et al. 2010). The critical resources identified by these owner-managers were financial and human resources. In all cases within the firms there were resource constraints so the issue became one of priority spending, budget allocation and human resource deployment. The firms in the sample suffered from declining revenues as a result of drastically reduced commission from suppliers and therefore had to make some key decisions about where limited resources were to be used. A recurring finding was that due to the simple decision-making processes and personal interests of the managers, priority areas reflected the personal preferences and interests of these owner-managers. Respondents stated very clearly that:

> "The priority is paying bills, paying staff and keeping the business running from day to day. I have seen nothing to suggest that my limited finances must be spent on something just because it is new. I don't think I need it."

> "I am not one of those who think that the use of the internet to reach clients is a waste of time but my profit margin is getting smaller and I have to prioritize how I use resources. My family depends on the survival of this business and I have to make sure that what I spend on will bring quick returns."

Resource-based theorists (Penrose, 1959; Chandler, 1977; Teece, 1982, Wernerfelt, 1984; Peteraf, 1993; Mowery et al. 1998) have overwhelmingly argued that those firms holding resources may maintain a relative position of competitiveness to other firms without those resources, once they act rationally. There did not appear to be a great disparity in the resource position of the firms, meaning that all firms stated that they were struggling in terms of revenues in similar ways because of commission cuts. The difference was not one of resource position but of perception between those technologically inclined firms and those which were not, as evidenced by the differences in the earlier responses of low adopters and high adopters. While there did not seem to be a resource advantage for these firms in the present, those firms which were acting to acquire greater online resources may hold such a resource position in the future.

The resource position barrier therefore provides a basis upon which firms already competing may outdo each other. Firms may therefore use their resource position to cement their lead. What is evident in these discussions is that resources play a significant role in a firm's decisions as it many times dictates what can or cannot be done. While this may be true in absolute terms, resources are never infinite even for larger firms with a strong asset base. It always comes down to the decisions which are made about how to allocate limited resources in the firm. With the exception of two firms in the sample, the interviews revealed that technology spending in particular on online platforms was not a priority area for the respondents. The other firms were mainly using GDS systems, internet for emailing and web browsing and did not aspire to climb higher on the technology hierarchy. Even those owner-managers who felt that there was some degree of importance to technological development stated that there were other areas of the organization which needed greater resources. These areas typically involved back office functions and paying staff. These areas of prioritization were based on three key issues in the firms.

The first issue is that these leaders had short-term vision and thought that it was not prudent to spend on anything that did not affect the immediate daily needs of the firms. This may be an indictment on the type of leadership being provided at the moment or more so of the leadership provided over the last 5-10 years. It may be the current short-sightedness that is causing the lack of strategic spending or the short-sightedness of the past may have placed these firms in survival mode now where markets and the global industry have already moved ahead of them. Firms in survival mode are unable to see beyond the immediate crisis with which they are faced and are many times spending resources playing “catch up”.

The second issue relates to the fact that these leaders were not convinced that they had received significant returns on previous technological investments. This may have translated into their perceptions that internet technology investments are high-risk, also a finding which was presented previously in this chapter. The third issue which relates more closely to competitiveness is that owner-managers were content not to spend on anything which it appeared their local competitors were not spending on. The fact that competitors were not spending their financial resources on cutting-edge technology was indicative of the fact that they were all given the GDS by suppliers free of charge. For the respondents this meant that they did not have to make major investments. Additionally due to the limited technology experience and low comfort levels with performing tasks online, it was their personal preference to use older, more manual approaches. Early resource theorists such as Peteraf (1993) more closely align resources with firm performance. This presupposes that firms are heterogeneous in terms of internal capabilities and resources, which has always been fundamental in strategic management discourse as it has always been argued (Andrews, 1971) that a firm’s strategies should emerge from an assessment of its organizational competencies and resources. On the basis of the heterogeneity of the firm it is therefore clear that strategies must differ between firms and must be dependent on what the firm is able and willing to do; a direct result of their asset base.

While it cannot be argued that the resource bases of the firms in the sample are identical, there did not seem to be sufficient heterogeneity to be able to identify a relationship to differences in firm performance. Their resource position in terms of asset base only affected performance to the extent that there was some degree of heterogeneity in their decisions about what areas to prioritize in terms of the budget. There was a more direct relationship between the leadership of the firms and the performance of these small, owner-managed firms. For example the firms, which turned out to be pioneers in the local industry in terms of distribution platform transformation were seen to have leaders with particular characteristics while those who lagged behind also had some leadership similarities.

The performance of the more innovative firms tended to be better in terms of sales, image and operational efficiency. It may be argued that this is coincidental; however the approach to leadership which influenced a greater openness to new innovative processes may also have previously translated into more proactive behaviours which have now led to these firms being market leaders.

Resource-based theories have also articulated that firms with broad resource bases are more likely to pursue diversification (Montgomery and Hariharan, 1991). This diversification may refer to diversification of products, markets or processes. The diversification of markets and processes in particular would be enhanced by a greater use of the internet for sales and marketing in these firms. While greater resources would enable firms to make decisions for example about what platforms they may or may not capitalize on, the decision to allocate resources to the introduction of advanced technology may act as a catalyst in the diversification of other areas of the business without further substantial investment. While a firm's resources certainly are not the sole determinant, they allow for a firm to have many more choices between strategy options as their internal capabilities allow for multiple avenues for conducting business. One exceptional firm in the sample demonstrated this concept and is now a market leader technologically as well as financially. The key driver which was identified was the difference in leadership perceptions about the role of technology in the daily operations of the business between the firms which were more inclined to adopt online technologies and those which were not. This will involve a broadening of their current understanding of resources to include technological resources and not just financial and human resources.

The inclusion of technological resources as a part of the asset base facilitates an extended discussion to include issues outside of the firm. Theorists (Teece, 1982; Mowery, Oxley and Silverman, 1998) have applied the resource-based view to inter-firm collaboration. Mowery et al. (1998) focus on technological capabilities and they posit that technological overlap plays a role in partner selection. The broader application however is that firms with significant resource bases are particularly attractive for collaboration. The literature does not point however to which category of resources are most attractive and also does not give much consideration to variances between industries.

These small, owner-managed firms in the travel industry for instance, do not see collaboration with their local competitors as a wise strategic decision. They base this on having to share their already limited resources. There was the sense that each firm felt that they were slightly better off financially than their competitors and therefore had something to protect from a strategic perspective. Respondents stated that:

> "Any collaboration with other travel agencies would hurt my business because we are all struggling and some of these firms are even worse off than mine which would mean that they would suck the remaining life out of my business."

> "Experiences in the past have proven that many of the other local travel agencies cannot be trusted. Any alliance would mean the sharing of certain kinds of information and from a strategy point of view I don't like it. Since we all have challenges with money then we will each have to figure out how to survive."

Resource-based discussions are grounded in debates about a firm's distinctive competencies, heterogeneous capabilities and diversification strategy (Mahoney and Pandian (1982). They further argue that distinctive competence is a function of the resources which a firm possesses at any point in time. The ability therefore to prolong or sustain this distinctiveness will depend on the type of resources and the type of industry. The RBV however is not intended to provide managerial prescriptions (Barney, 2001). While this is so, it points very clearly to implications of why some firms have a strategic competitive advantage over others and can therefore be of great value to managers (Kraajenbrink, et al., 2010). Most of the firms in the sample do not have a distinct asset advantage. It was noticed that strategies varied to some degree with high adopters and low adopters and a relationship between strategy and resources was identified. High adopters to include two firms that were exploring online selling options, one with a significant asset base and one with a limited asset base, tended to have proactive strategies which recognized that they should create platforms which anticipate changes in the market. They therefore were willing to allocate resources to support those strategies. In Moderate and Low adopter firms (*see Fig 5-7*) they were reactive and typically stated that "local markets were not yet ready for this type of online activity."

Contemporary Resource Theorists (Brouthers et al., 2008) have extended discussions of the resource based advantages to the international context. They suggest that differences in the institutional environments of nations may influence its applicability. Therefore the resource based perspective may become more robust when the moderating influence of national institutional environment is also considered. Context specificity may therefore play a role in assessing firm performance. Evidence from the field demonstrated that all of the firms in the study operate within the same national institutional environment, thus differences in the spending choices clearly stem from separate factors which are unique to firms. This however does not preclude the application of the RBV but rather suggests that firm-specific capabilities and resources be considered within broader institutional contexts and firm strategy considerations for studies of a more comparative nature. A more useful analysis of the impact of resources in these firms would surround how competencies are enhanced or constrained by resource capabilities.

The firms emphasized in the interviews that there were significant limitations placed on these small firms due to a lack of sufficient resources. "You have to understand that we are a small company and we don't have a lot of money to go around." While this was a recurring theme throughout the data collection process, it was more important to establish whether there were significant differences in resource bases across firms. The heterogeneous nature of firms based on resources has received substantial attention in the literature. In addition there have been discussions about the competitive edge that these differences provide. It has already been established that there is greater homogeneity than heterogeneity in terms of resource bases in these firms. There was a difficulty in attaining specific data about financial resources of these firms. However all firms stated that they are operating at margins which were below their variable costs, which is usually an indication from an accounting perspective of whether firms should stay in business.

There was more heterogeneity between how firms chose to prioritize the use of their limited resources. This was a direct result of leadership decisions, which as stated earlier in this chapter, are paramount due to the small size and simple management structure of these firms. The actions of the leader become important in decision-making regarding resource allocation (Kraajenbrink et al., 2010). This is even more magnified where firms have significantly limited resources as tough decisions must be made about priority spending in order to efficiently use these resources. The resource limitations which were highlighted by owner-managers focused primarily on human and financial resources. In most cases the managers stated that there was not enough money to invest in a hi-tech strategy. In other cases however managers felt that where money was available they did not have capable human resources to consistently maintain technological innovations such as websites. This was also evidenced where firms in the minority which had websites stated that these were inactive. Although these resource limitations existed it was evident that technology adoption and training were not seen as priority investment areas based on leadership resistance. The respondents highlighted that with the limited financial and human resources they were more inclined to use resources in areas that would assist in the daily operations rather than on strategic initiatives which they were not sure would yield significant returns. They had to focus on what was important now in preventing the termination of their business operations. Technological innovations were mentioned as being risky investments which were more geared towards the long-term and many respondents argued that it was not considered a priority area since it did not address the immediate needs of the markets being served. This highlights a transactional approach to leadership, which simply seeks to enforce procedures and control daily activities.

The situation for the firms under investigation is also a unique one in that the respondents are owner-managers who have a stake in the immediate or short-term benefits of the operations. This work posits that owner-managers in a stand-alone environment are inherently transactional leaders based on the challenges of the industry coupled with their desire to realize immediate returns on investment as this feeds directly into their family finances. This is an extension of the entrepreneurship literature and particularly the work of Covin and Slevin, (1988) which argues that small business managers prefer to be decidedly risk-averse, non-innovative, and passive or reactive. Carland et al, (1988) state that they operate their business as an extension of their individual personality and immediate needs. In addition to a reluctance to adopt internet technologies, this is evidenced by the sample's refusal to engage in inbound travel due to a lack of commission although this may yield long term benefits.

Other firms in the sample which either had numerous branches, had established brands or which were affiliated with other business in other industries tended to behave differently. It appears then that the process of prioritization for resource allocation is influenced by owner-manager's personal needs, preferences and short-term fatalistic goals, which do not consider a strategic vision for these firms. In attempting to further explore these issues, respondents were asked to assess their current technological needs and match these to financial and human resource capabilities. For this aspect of the discussion the major technological focus was on websites since most firms were just below that level on the hierarchy of adoption. Additionally carefully designed websites would allow for online sales activity, which is ultimately being examined in this study.

Overall the employees who worked in these firms had low levels of formal education. The typical worker only had a high school certificate and 3 GCE O'level subjects or the equivalent Caribbean Secondary Examination Certificate. This is lower than other front line employee jobs such as banking which at the very minimum required 5 subjects and more recently preferred to recruit candidates with Bachelor degrees. When asked to describe employees' experience with learning to use new technologies, respondents stated that most of the staff usually found it very difficult. Most firms did not have a budget line item for general staff straining so it was not surprising that technology training was even less evident. The dilemma seems to be that firms are hiring the cheapest labour and then refusing to add value to employees through training. There is already a resistance on the part of many of the owner-managers to the adoption of some types of innovations and they are further discouraged by the prospect of having to provide training for employees who are likely to be difficult to train. There is also an added dimension in that the types of employees in these firms are less likely to be creative thinkers and may simply enjoy routine work. The implication is that even where managers are intellectually stimulating, they would receive very little response from employees who are more content to follow directives. Many owner-managers argued that all of their human resource considerations such as recruitment, selection and training were directly related to their financial resources. Although a few respondents still stated that a greater financial resource base would not guarantee greater technology spending in the short to medium term, all of the firms would allocate greater spending for the training of their human resources. An examination of the financial resource base will now be carried out.

The declaration by owner-managers that financial resource constraints are the deterrent to technology adoption in their firms is incongruent with a subsequent finding. Most of the sample said there was no other technology in which they wanted to invest and could not do so because of a lack of financial resources. The suggestion that the lack of resources is a barrier would imply that there was a desire to adopt but this was being hindered by the lack of resource capacity. This highlights a general unwillingness to adopt more than an inability to adopt. Having established the previous point it is still important to make mention of the financial challenges of these firms. While it does not directly drive adoption decisions, there is an indirect influence. This is found in the fact that the weak financial positions of the firms keep them in crisis management mode with little time for serious strategic planning. This is particularly poignant for owner-managers of these small firms whose livelihood depends on the survival of the business and its profitability. This presupposes that the over-arching driver emerges as leadership and the variables surrounding each leader, which makes him/her different from their counterparts in the local industry.

An exploration of internal factors such as strategy and resources was necessitated by previous works which suggested that a firm's resource capabilities (Barrney, 2001; Kraajenbrink et al. 2010) and its strategy decisions (David, 2007; Stonehouse and Snowdon, 2007) affect its behaviour. The behaviour being investigated is technology adoption behaviour for small, owner-managed firms particularly for online sales and marketing activities. Multiple factors, which affect decision-making in firms such as strategy, resources, leadership, culture and the digital divide, were explored through secondary and primary data. The analysis reveals that throughout the process of distillation to identify the key driver of technology adoption for small owner-managed firms, strategy and resources were not the central factors. They should not be dismissed however as they are peripheral contributors in the sense that they are used by leaders to affect technology adoption processes but they do not provide the genesis for these processes. In many ways they emerge as a result of leadership behaviours rather than drivers in and of themselves. They may turn out to be more influential in contexts which are different but for the purposes of small, owner-managed firms in information intensive industries they are more pawns in technology decisions rather than key determinants.

Firm strategy and resources were identified in the literature as being important elements of the discourse on firm behaviour as demonstrated in *Fig 1-1* These factors along with leadership, culture and the digital divide were also framed within the theory of organizational decision-making in *Fig 1-2*. This comprehensive approach was taken to identify the drivers of technology adoption behaviour in small owner-managed travel firms along a technology adoption hierarchy. The further exploration of these internal factors (strategy and resources) through primary data collection with all owner-managed travel firms in Jamaica revealed that there was no clear strategy which influenced technology adoption behaviour. It could be inferred that many firms essentially engaged in a hi-touch strategy (intense personal interaction) and feared that they could suffer from a perceived psychological risk of dependency on technology as posited by Eastlick and Lotz, (1999). Additionally owner-managers felt that a hi-tech and hi-touch strategy were incompatible.

From a resource perspective owner-managers indicated that they face human and financial resource constraints. Their claim that this inhibited their ability to adopt greater levels of technology was anachronistic with a later finding in the interviews that they did not feel that there were other technologies which they needed which were unattainable because of resources. The implications are that the decisions of how to allocate already limited resources are the sole purview of owner-managers. The tendency to prioritize spending in areas other than technology adoption was a reflection of the individual leader's preferences and comfort zone.

In the final analysis firm strategy and resources did not sufficiently explain differences in adoption behaviour in the firms. These factors were useful in an understanding of what tools were used to drive technology decisions in an overall sense but were less significant in explaining why some firms were at different levels of the technology adoption hierarchy as strategies and resource constraints were not heterogeneous enough to answer the question of why some firms were more technologically advanced than others. These internal factors emerged as a product of differences in the leadership of these small, owner-managed travel firms rather than as significant input factors into the decision to adopt cutting - edge technologies. In light of the failure of the internal, micro factors to explain adoption behaviours for these firms, it is imperative to investigate and discuss whether more macro factors such as culture and the digital divide may explain firm adoption behaviour.

CHAPTER THREE

EXTERNAL FIRM CHARACTERISTICS: BRIDGING THE DIGITAL DIVIDE

In the Organizational Theory literature, culture is conceptualized as either national culture or organizational culture. For the purpose of this study the national culture element will be the focus. This is due to the fact that small, owner-managed firms are being analyzed and it has already been established in the previous findings chapter that factors which are internal to the organization are directly controlled and manipulated by these leaders who have a vested interest in their businesses and often project their likes and dislikes into business practices. With this in mind internal factors such as organizational culture, structure, strategy and resources are relegated to being secondary drivers and may be considered to be more of an effect than a cause in understanding firm behaviour in small owner-managed firms. National culture as an external factor will be conceptualized in terms of values, norms, traditions and social interactions within the society.

Since the seminal work of Hofstede (1980) which identified international cultural differences in the workplace, many authors across a range of fields have relied on this work. Apart from the critique that there was an over-dependence on Hofstede's work, the most resonant critique emerged when McSweeney (2002) argued that any work that assumes uniformity within national cultures is flawed because there is significant observable plurality within nations that produces differences in behaviours. He questions whether culture can systematically cause differences in behaviour between people from different countries. The research findings reveal that there were observable differences in technology adoption behaviour across firms although they shared the same national culture. This prompted an exploration of specific cultural constructs which influence behaviour and more specifically adoption behaviour in firms.

Societal values played a role from the perspective of the customer. Owner-managers typically stated that "the most important and valuable thing to customers is the personal touch." For many respondents this meant that they should concentrate their efforts on providing exceptional one to one experiences rather than seeking for the most efficient distribution method. Also the fact that they viewed online sales as holding very little value for their customers led them to believe that the introduction of some innovations was pointless. One respondent stated: "We could use all of these fancy technologies but if our customers don't see it as something that is necessary then we would all be wasting our time trying to do something just because it's fashionable."

The values of the customers influenced the values within the firms to a large extent, and the previous response is representative of the majority of the firms. There were differences with high adopters who were exploring online transactions and social media utilization. The two firms at the top of that hierarchy held divergent views from low adopters and stated that:

> "While a large percentage of the Jamaican population does not place a high value on e-commerce activities today we recognize that those values are changing fast and we can even see this in the wide acceptance and use of smart phones such as blackberry."

> "We have to be prepared for market changes which are on the horizon. Even if we only continue to serve this local market, there is a revolution taking place with our young people which we must anticipate."

It is useful to highlight these differences since all of the respondents have been exposed to similar national value systems. The differences in their outlook suggests that there are other key factors outside of the value construct which contribute to different adoption behaviours among firms. These differences are still useful for academic consideration to the extent that uniformed differences in firm behaviour are identified within specific national cultures. Some authors (Gelade et al. 2006) continue to explore the impact of national culture on the organization and identify that there are significant relationships between organizational behaviour and some aspects of national culture such as norms and traditions.

The norms and traditions of the society which represent the way in which the people are accustomed to seeing and doing things played a role in overall cultural influence. While this could be seen in the local client base, it was more evident within the firms. Respondents articulated that: "We have been doing it this way for over 20 years and it has worked for us. If we change our approach we do not know if it will work just as well. Also most of us in this company are not comfortable with changing how we make bookings." Coupled with being steeped in traditionalist ways of doing business, the respondents seems to suffer from what Lang (2000) referred to as "techno fear". The two are not separate in the sense that it is a subscription to norms, customs or traditions that necessitate a rejection of anything that goes against the usual ways of doing things. A very critical finding was the fact that high adopters had different views in relation to traditions. The leader of the most technologically inclined firm stated that:

"Nothing remains the same forever. There was a time when we did manual bookings and we did not have any GDS. We learned to use the GDS and today many of us cannot do without it. In the same way we always have to be ready to adjust based on new developments." The divergent views once again point to the fact that while national norms and traditions play a role in overall societal views, there are other factors which must explain the varying perceptions and adoption behaviour from one firm to another. It is of vital importance to highlight that all of the owner-managers in the sample have lived for their entire lives in Jamaica, so outside cultural influences are somewhat limited apart from media exposure and short stay travel.

An indirect cultural influence emerged as owner-managers highlighted that the culture shaped the customers and therefore their business practices had to respond to the market preferences of a relationship-oriented society. This relationship orientation is symptomatic of the perception that customers in the local outbound market being served place a high value on personal interaction which could not be replicated online. This points to issues of marketing and distribution strategy but also emphasizes a constraint on these leaders' ability to make some decisions. Respondents articulated that:

> "I would love to see the internet listen to their complaints and empathize and ask how their family is doing. These are the things which my customers love because Jamaicans are very friendly people and I have a good relationship will all of my customers."

> "A lot of my business comes through a major company that I used to work for because I still have a good relationship with them and their families. They always book with me no matter what and we are good friends even before we are service provider and customer."

> "Someone in Jamaica always knows someone else. Once you form a relationship with a client they always tell someone else how warm and friendly you are. Before you know it you have a new client because they recommended you and that person wants the same personalized treatment."

The relationship orientation of the society influences technology adoption in the firms in a negative way as most respondents were confident that serving a relationship oriented society where personal interaction was much more important than speed and efficiency, would secure their business in the future. High adopters were less inclined to take this stance. The market leader pointed out that:

> "A number of things are changing. First of all our loyal customers are not as loyal anymore, so the relationships seem to mean less to them now. Secondly these loyal clients are getting old and are traveling less so we have to focus on the new generation and they are all about the best price and convenience. And don't forget that they are internet savvy. As I mentioned to you before they do everything on their mobile smart phones."

Similar to the previous constructs of values, norms and traditions, the relationship oriented culture plays a role for firms which are low on the adoption hierarchy but does not explain why the minority of firms deviate from these norms and seek to become proactive in engaging in innovative behaviour. Understanding differences in innovative behaviour warrants a closer look at the culture and innovation relationship. The primary data collection and analysis revealed that national cultural background generally shaped the views of individuals about technological innovations based on the admission by leaders that they were influenced by what they observed in society. This meant that relatively low levels of observable technological activity in the country resulted in an inability of respondents to see the technology in action along with its benefits. One respondent stated that:

> "All of these things that you mention are things that I hardly see anyone use. I don't think that Jamaica is a country that uses a lot of technology. I know that the latest craze is Facebook but young people only use it to talk to friends. I can't say that I have seen any evidence to suggest that we are high technology users."

The challenge is that this does not explain why even with a shared national culture there were different adoption behaviours among the firms in the sample. A deeper exploration and analysis of the data revealed that while national culture did not explain these differences, dominant sub-cultures to which these individuals belonged played a key role in shaping their perceptions about technology use. The family emerged as the most significant subculture in influencing perceptions.

Sub-cultures have their own values, norms and traditions. The family is one such sub-culture and provides a primary source of socialization (Haralambos and Holborn, 1980). Individuals within a society are therefore influenced to engage in some behaviours more by their sub-culture than by the broader national culture. In light of the fact that leadership has been established previously in this research, as a critical factor in determining what new technologies to adopt in the case of small, owner-managed firms, it is important to identify what factors shape the development of these leaders. From a cultural perspective the family background of respondents played a role to the extent that there was innovative behaviour taking place in the families. Leaders who came from families which had children who were engaged in technologically savvy behaviour for example were more inclined to engage in those behaviours personally and as a result project these activities into their businesses. One owner-manager pointed out that "I am very open to using the technology personally because I have seen my son buy all these things online using my credit card ofcourse. As a teenager he has taught me how to buy things online too. I realize that if I can buy online then why not sell online. This is why I am now exploring the option." This stemmed from a greater level of comfort with using tools which they had observed a trusted member of the family utilizing.

Culture and Leadership

In support of the work of Elenkov and Manev (2005) this work has found that cultural context influences leadership and moderates its relationship with organizational innovation. There is some dissimilarity however in the importance placed on cultural influence as well as the point of emphasis. This research has found that the influence is more indirect than previously posited by the above authors. While a cultural influence has been recognized in the sample, there were some key differences in behaviour which could not be explained by national culture since owner-managers all had this is common. A comparative study with European (Elenkov and Manev, 2005) or Asian (Jung et al., 2002) countries may reveal some differences as a result of national culture but within the same culture there is less resonance.

Additionally the discovery that the most influential subculture in influencing innovative behaviour was the family was insufficient to completely explain adoption behaviour in the firms. There were some firms at higher levels of adoption which had a similar family composition to some firms at a lower level of adoption. Although the evidence showed earlier that leaders agreed that the family influenced their comfort level with introducing some technologies in their firms, there were still differences in levels of adoption. An overall influence could be seen in the differences between high adopters and low adopters but this was not as distinct between moderate adopters and high adopters who had family composition in common. This meant that the differences in firm behaviour needed to be explained alternatively through an exploration of other factors, which in combination with family, could explain leadership differences and ultimately technology adoption differences.

The contemporary work of Singh and Krishnan (2007) highlighted that transactional and transformational leadership must however be treated and addressed with more cultural sensitivity in the leadership literature. With this in mind it is important to justify a relationship between culture and the development of leaders and to also identify that although leadership is seen as the key driver in technology adoption in this research, there are broader considerations from a *Critical Social Science* perspective which may play a role in the process such as the family.

It was interesting that respondents highlighted that the national culture shaped their local competitors views of technology which therefore reduced the need for them to adopt some technologies since the competition was not doing so. Each respondent stated this about the other respondents. It appears therefore that while they do not perceive a cultural influence on their own leadership style or propensity to adopt innovations, their counterparts overwhelmingly state that the converse is true. The challenge is that this does not explain two things. The first is whether culture impacts on leadership characteristics as defined in the transformational leadership literature or whether it merely influences some views of technology. This is an important distinction since strong transformational characteristics should cause the leader be able to abandon pre-existing views where there is sufficient evidence of the need for change. Secondly it does not explain why there are different views about technology within the same culture. The cultural influence is therefore treated as an indirect one which affects leadership peripherally and does so in conjunction with other factors.

The understanding of culture is related to a broader sociological discourse which identifies norms, values and customs as being important constructs in any discussion of a social system. It must be noted here that a number of the classical theorists of technology adoption such as Rogers (1962) and Davis (1989) identified the important influence of the social system on adoption. Each social system has unique characteristics that make it different from other social systems. These cultural differences affect work related values, behaviour and innovation. Ultimately a leader was seen to be the product of cultural orientation. These conceptualizations warranted an exploration of culture as a determinant; however it was found in the primary research that the influence is only indirect. While a cultural influence has been recognized in the sample, there were some key differences in behaviour which could not be explained by national culture since owner-managers all had this is common. Additionally the influence of the family as a subculture explained larger disparities but did not offer an explanation for subtle differences between leadership categories. This necessitated an examination of other factors to identify explanatory variables. The digital divide may provide a deeper discussion especially when conceptualized globally as well as within societies. Any observable inequalities particularly within the society may provide an explanation of differences in firm behaviour regarding technology adoption.

The Digital Divide

A new entrant to the Jamaican communications market, Flow Jamaica, recently laid a new submarine cable connecting Jamaica to the United States. This new cable increases the total number of submarine cables connecting Jamaica to the rest of the world to four. Although broadband penetration was slower than countries such as the United States and the United Kingdom, levels of connectivity were still high through the use of smart phones. This supports the work on mobile technologies by Stump et al. (2008) which identifies that this has been diffused rapidly throughout developing countries which has in many ways lessened the disparity between nations. A global digital divide cannot be ignored however as the average Jamaican home still does not have broadband connection even with increased technology infrastructure. One respondent highlighted this by stating that:

> "It would be a waste of time to try to reach our customers online because many of them do not even have the internet at home. Even though the younger ones are using smart phones most of them are not buying anything online with their phones at all."

Respondents highlighted that they felt that there was sufficient technology infrastructure provided at a reasonable cost, however in their view many Jamaicans were not interested. "The technology is available but I just don't think we are that kind of society. We thrive on personal contact." This is in keeping with the annual e-Readiness survey routinely conducted, compiled and published by the Economist Intelligence Unit (EIU), which is the business information arm of the Economist Group; publisher of The Economist magazine. In Jamaica's third showing in the 2007 EIU report there was a marginal increase in Jamaica's overall score from 5.03 to 5.05. However, there was a fall in the ranking of three places from 43 to 46.

There is now a greater focus in the survey on broadband affordability and penetration. Jamaica had previously ranked well in the area of telephone and narrow band penetration, however with the removal of that category the performance fell in the area of Connectivity and Technology Infrastructure. Jamaica's score also fell in the category of Consumer and Business Adoption as a result of a re-vamping of the criteria in this category; moving away from the availability of finance and IT personnel, to the number and level of online commerce and government services. Online commerce for travel services in particular is still virtually non-existent and it is an area which needs to be addressed. This is however not an issue of access, therefore the access divide posited by theorists above does not explain adoption levels in firms. As the conceptualization of the digital divide in chapter 1 explains there are also situations of unequal access within societies. The evolution of theory on the digital divide reveals that these may stem from socially driven inequality (Lash, 1994), politically driven inequality (Feather, 1998) and economically driven inequality (Yu, 2006). The data revealed however that while there were perceptions that there was economically driven inequality in particular, this did not separate firms which were high technology adopters from those which were low technology adopters. Overall the access to technological resources, though less than in developed countries, was equal within the business context for Jamaican firms. Arguments put forward by James (2004) articulate that there is a closing of the divide based on access. He argues that although individual access can be considered limited in some of these countries, there is still innovative use of technologies in some poorer developing countries. It however does not challenge the position of Rogers (2003) which states that while the access-divide has received most of the attention, it is the learning-divide and content-divide among others which will present a disadvantage for some. This argument provides key insights into issues of matching content to audience needs.

The content and learning divides provided an explanation for differences in firm behaviour in general but these were also related to other factors. For example it was found in the interviews that there were slight differences in terms of the learning divide but this related to the skills and competencies of the employees. This is also a human resource issue and fits more directly into the resource-based discussion. The content divide did not play much of a role as most of the content being developed was created with western culture preferences in mind and the Caribbean benefitted from being in such close proximity to the United States. Most importantly notions of the digital divide must be applicable to industries. The general argument put forward by proponents of the global digital divide (Mosaic Group 1998; Castells 2000; Norris 2000;Rogers 2003; Kirkman et al. 2002; Drori and Jang, 2003) is that economically developed countries control ICT development and access and therefore automatically have a distinct advantage. The major criticism of this body of work is that there is very little specification by way of solutions (James, 2004). While much of the earlier research has been concerned with the information inequality within societies, there has been more recent focus on the global digital divide which addresses differences across countries. This shift in focus has encouraged a more robust debate in the travel and tourism context where demand and supply issues are addressed. Indeed maybe the solution specificity being sought may emerge from specific industry research such as travel and tourism.

Maurer and Lutz (2011) support the work of Minghetti and Buhalis (2010) and argue that there is a significant communication gap between supply and demand side elements of the travel and tourism industry. They posit that there are four types of communication gaps which exist: access gap, usage gap, skills gap and attitude gap. They take the position that these are hierarchical in nature and that one level has to be complete before the other.

While this does not explain differences in firm behaviour in the study it was important to analyze these issues on the grounds that there are industry implications. There is a gap which may be filled by these retail travel firms whose importance has decreased drastically in the last two years. The National Tourism Office, The Jamaica Tourism Board has been unsuccessful in closing the digital communication gap with developed countries, which presents an opportunity for these outbound agents to become more engaged in inbound activity by increasing the online interaction between countries. If this re-engineering of their functions is to take place there needs to be an understanding of current gaps which exist. The interviews revealed that the gaps between the firms in the study and developed markets which demand tourism services include each of the four areas earlier espoused:access gap, usage gap, skills gap and attitude gap. This work did not find that this hierarchical sequence was as uniformed as previously advocated by Maurer and Lutz (2011).

The data revealed that while access was a precondition for usage, the same was not true for the upper tiers of the hierarchy. In fact in many of the cases an attitude gap and skills gap existed concurrently and in other cases the skills gap appeared to be higher on the hierarchy than an attitude gap. The findings of this research run counter to the argument that attitude is always at the top of the multiple digital divide pyramid. For the purposes of this work the digital divide was conceptualized in terms of whether there was information inequality as driven by economic, political and social factors. Economic factors were most dominant as it related to access to technological resources when compared globally. This gap was now being significantly reduced as Jamaica experienced phenomenal levels of mobile smart phone penetration. There is still a communication gap between the destination and developed high-access countries from which the country gets its international tourists.

More importantly, an assessment within the society revealed equal access to technological resources for these small, owner-managed firms. This means that the digital divide clearly does not provide an explanation for differences in firm behaviour within the society. It is only useful in this study to the extent that it informs overall industry challenges for the firms in the study. In the face of these challenges which are common to all of the firms in the sample, firms still behave differently in particular in terms of technology adoption and these differences require alternative explanations. Context issues such as the digital divide and culture have an overall impact in explaining the slow IT uptake of the industry, however given that there was no evidence of access inequality or significantly different cultural values between the firms, they could not inform adoption variances between the firms.

It is evident that contextual issues such as culture and the digital divide may explain overall industry behaviour but they are inadequate in explaining differences in behaviour among firms. The firms being investigated are faced with similar environments in which to operate and yet there are key differences in the operational and strategic choices which they make. The examination of culture as a factor revealed that firms were operating in a relationship oriented society where friendships and familiarity influenced business relationships. In many cases firms gained clients from the word of mouth of existing clients who became "friends of the business". This had an effect on views of whether increased technology adoption was necessary in two ways. Firstly some respondents felt that their client base was always secure and stable and that they did not have to do much to encourage their loyalty. Secondly they stated that they knew their clients extremely well and argued that their relationship orientation meant that they were culturally predisposed to the need for intense personal interaction, which could not be provided by the internet.

In addition to a relationship orientation, the national culture had an impact in that it promoted traditionalism with strict adherence to norms within the society. This had implications for firm behaviour as well as customer behaviour and influenced strategic decisions which required a deviation from these norms. These cultural factors at the national level were experienced by all members of the sample which indicates that they may not be used to explain individual differences in firm adoption behaviour. A more determining element of culture stemmed from the sub-cultural influence of the family.

Family composition played a role in firm technology adoption through its influence on leaders of these firms. While the family background of employees may also have shaped their views it was deemed to be more important to assess its influence on leadership in the context of small owner-managed firms where leaders make unilateral decisions about innovation. It was found that family composition played a role especially in separating high adopters (those in the initiation phase of online selling and the use of social media as a marketing tool) from those at the bottom of the adoption hierarchy. Those owner-managers who had family members living in the same household who were engaging in innovative behaviour were more open to innovative ideas and were more likely to introduce innovation in the firms. In particular active online shopping teenagers made a big difference as respondents were able to see the benefits of operating in cyber space. Although the influence of the family as a subculture explained larger disparities it did not offer an explanation for subtle differences between leadership categories. For example there were moderate adopters and high adopters who had family composition in common. This means that this variable is not explanatory in isolation and must be used in combination with other variables to explain variations in the technology behavior of firms. An even less significant context factor in explaining firm adoption behaviour was the digital divide. The firms in the sample had equal access to technological resources and faced similar challenges in terms of the content and learning divides. The socially, politically and economically driven information inequality within the society related more to the wider population than to these middle class business owners.

The digital divide debate is more useful when applied to the travel and tourism industry as a destination imperative rather than as a firm imperative. There is more of a global divide than one which is internal to the country in the business context. These small owner-managed travel firms may reintroduce themselves as being relevant through bridging the technological divide (Minghetti and Buhalis, 2010) and communication gap (Maurer and Lutz, 2011) which exists through the use of covertly induced and autonomous agents such as the internet (Govers et al. 2007). Given that a level playing field has been identified in terms of national culture and the digital divide within the country, the explanatory potential of these factors has been weakened in understanding technology adoption differences in the firms. The factor which has emerged as most significant in explaining why small owner-managed firms with equal access and opportunity to technological innovation make choices which place them at different levels of adoption is leadership.

It has been shown that leadership is the key determinant of technology adoption behaviour in small, owner-managed firms. Firm strategy and resource allocation emerged as factors which were manipulated to meet the needs of owner-managers and were more of an effect than a causal element in the technology adoption discourse for firms of this nature. Broader macro factors such as culture and the digital divide provided more explanation of why the travel and tourism industry as a whole was slower on IT uptake than industries in other countries. Due to the fairly even influence of these factors on the firms within the society, they are instructive only to the extent that they can provide an enlightened approach to industry restructuring. The aim of this study is primarily to determine what factors influence the adoption decisions of small owner-managed travel firms and the key factor which influences this is leadership.

These firms are seen as extensions of their individual leaders which are the owners of these firms and make all decisions whether minor or major. These decisions typically reflect their individual preferences and experiences. The following chapter provides a detailed discussion of the leadership element and what variables influence a greater propensity to adopt technological innovations in these firms. Ultimately these leadership characteristics coalesce into leadership typologies which influence staged adoption for the firms along a hierarchy.

CHAPTER FOUR

TRANSACTIONAL VS. TRANSFORMATIONAL: LEADERSHIP CHARACTERISTICS FOR TECHNOLOGY ADOPTION

Leadership has emerged as the most significant factor driving technology adoption for sales and marketing in owner-managed small travel firms. While it may be argued that leadership is influential in many different types of firms (Hitt et al., 2001), the key characteristics which make this construct more resonant here are size and owner-management. Smaller firms may mean a more simplified hierarchy and less bureaucracy in decision-making than larger firms (Brown et al., 2007) which invariably places more control in the hand of a single leadership figure. The decision-making power of a leader is also magnified where owners are themselves the managers of day to day operations. In the early work of Carland et al, (1988) they state that these types of leaders operate their business as an extension of their individual personality and immediate needs. This suggests that decisions which affect firm behaviour are likely to be a reflection of the leader's preferences and personal situations.

This research does not attribute sales and marketing technology adoption such as the internet solely to leadership, as it recognized the influence of other factors such as culture, strategy and resources in previous chapters. However, the two previously mentioned factors of size and owner-manager influence, make the leadership factor the most powerful driver of technology adoption in these firms being studied.

The evaluation of the findings from individual variables produced some key variables of education, previous work experience, technology experience, risk aversion, family composition and intellectual stimulation. These ultimately led to the conceptualization of leadership categories termed as resistors, caretakers, stabilizers, reactors and transformers, which apply more directly to the technology adoption discourse. These typologies are formed from various combinations and degrees of the aforementioned variables. Each variable and typology will be discussed in detail in the subsequent chapter but will be alluded to here. Prior to this, a detailed breakdown of all the relationships explored will be presented in order to show how these variables emerged as the critical ones in explaining each new leadership typology.

Leadership Background

The background of the owner-managers in the firms which were interviewed revealed remarkable diversity especially in relation to education. It was found that the formal education levels were fairly low.

Table 4-1: Formal Education Levels

Education Level	Frequency	Percentage	Cumulative Percent
Secondary	6	19.4	19.4
Certificates/Diploma	11	35.5	55
Bachelors	9	29.0	93.5
Masters	5	16.1	100.0
Total	31	100.0	

As seen in *Table 4-1,* more than 55% of these owner-managers did not have a university degree. Typically the highest levels of qualification were certificates or diplomas after secondary school which would have been issued by professional travel bodies rather than established universities. On its own this did not provide much explanatory potential so this was further investigated through a cross-tabulation of the relationship between qualifications and perceptions of internet use in sales and marketing. While the sample size does not allow for a statistically significant claim to be made, this exploration is useful in that it identifies relationships for further qualitative analysis.

Table 4-2: Cross-Tabulation: Highest Qualification and Perceptions of Internet importance in Sales and Marketing

Highest Qualification	Very Important	Important	Unimportant	Very Unimportant	Total
Secondary	0 0%	0 0%	5 83.3%	1 16.7%	6 100%
Diploma	0 0%	6 54.5%	5 45.5%	0 0%	11 100%
Bachelor	2 22.2%	6 66.7%	1 11.1%	0 0%	9 100%
Masters	3 60%	2 40%	0 0%	0 0%	5 100%
Total	5	14	11	1	31

Higher levels of education generally had a positive correlation with perceptions of the importance of the internet in sales and marketing. *Table 4-2* demonstrates that the only group where all of the respondents stated that the internet was unimportant or very unimportant was the group where the highest level of education was secondary school. The perception of importance gradually increased with each level of education and ultimately the only two groups where the internet was viewed as very important in sales and marketing were those groups with university educated respondents. A similar pattern was also observed in *Table 4-3* where cross-tabulations of education level and perceptions of the importance of the online market to the business were carried out.

While there was no group of respondents which viewed the online market as being very important, the only group where all of the respondents felt that this market was important was the group of masters educated owner-managers. Not surprisingly, those leaders with university qualifications were more open to the idea of the use of various types of technology due to greater exposure to technology use and formal discussions of its benefits in a classroom environment. One respondent highlighted:

> "In a university setting you are forced to use various forms of technology. For instance we had to use the internet to communicate and we even had to write papers on how technology has revolutionized the business world. There are some obvious benefits but there are also some costs."

Further exploration revealed that those respondents who had a bachelor and masters degree as their highest qualification predominantly held qualifications in business and management while some had their training in travel and tourism. The most interesting finding here was the fact that those who had more general business degrees rather than tourism specific ones were more inclined to use online platforms for sales and marketing. This may be due in part to a greater emphasis on competitive strategies and the development of sound business models in business studies. This finding was gleaned from open-ended responses about educational backgrounds and doing a comparison with technology perceptions. This has provided an exploratory basis for future research on education type and leadership.

Table 4-3: Cross-Tabulation: Highest Qualification and Perceptions of Online Market Importance

Highest Qualification	Very Important	Important	Unimportant	Very Unimportant	Total
Secondary	0 0%	0 0%	0 0%	6 100%	6 100%
Diploma	0 0%	0 0%	11 100%	0 0%	11 100%
Bachelor	0 0%	0 0%	9 55.6%	0 0%	9 100%
Masters	0 0%	4 80%	1 0%	0 0%	5 100%
Total	0	4	21	6	31

Deeper qualitative enquiry highlighted the study interests of those respondents who did not have university degrees. All of these owner-managers expressed an interest in receiving more formalized training in travel and tourism. Reasons given were: for a greater understanding of how the industry works as a whole, understanding new and emerging markets and strategy development. The indication that these areas need to be strengthened highlights weaknesses in the management information intelligence of these firms. Having been trained to certificate levels, these owner-managers felt that this information intelligence gap could be addressed by further travel and tourism training, which may represent a tunnel-vision approach to understanding how businesses operate and function in a globally competitive environment. The knowledge gap is partially related to education and training. Another equally important variable which emerged in the interviews was that of experience using various forms of information and communication technology. It was felt that the personal experience of these influential owner-managers within firms with using the internet may provide insights into their propensity to introduce it as a sales and marketing tool in their firms.

Table 4-4 Frequency of Internet Use (Leaders)

Average Rate of Usage	Frequency	Percentage	Cumulative Percent
Once per week (Emailing)	5	16.1	16.1
Twice per week (Emailing)	12	38.7	54.8
Every other day (Emailing and web browsing)	10	32.3	87.1
Every day (Emailing and web browsing)	2	6.5%	93.5
Every day (emailing, web browsing and online purchasing)	2	6.5%	100
Total	31	100.0	

The rate of use for owner-managers in the sample was quite low. More than 50% of the respondents only used the internet twice per week and just 13% used it every day. The rate of use however tells very little on its own and so an understanding of the reasons for which it is used will add more robustness to explanations which will emerge from these findings. Those respondents who used the internet on average once or twice per week did so to check email and clear their inboxes while respondents who were using the internet every other day were checking email and browsing the web for information. None of these categories of respondents had ever done any online financial transaction. Of the four (4) respondents who used the internet every day, two (2) of those respondents engaged in the same activities of emailing and web browsing, while the other 2 were emailing, browsing and using personal Facebook accounts on blackberries. It must be highlighted that the more innovative of these two (2) respondents had made a personal online purchase in the past. One owner-manager articulated that this was influenced by seeing his son make successful online purchases. He stated: "At first I was very wary of doing certain things on the net until I saw my 16 year old son make it look so easy. He was buying music and gadgets online and even though it took me a while I eventually tried it one day and the rest is history."

The personal internet uses of the leaders of these firms translated into business practices and firm adoption behaviour. The only firm which had engaged in online sales was one of the two firms which had internet savvy leaders. The other had intentions to develop a better online platform for sales as their inactive website was an insufficient one. However personal Facebook use did not translate into firm use as they did not view social media as a business tool. The firms whose leaders engage in minimal internet use for emailing up to two times per week experienced low levels of internet involvement and primarily used the web for email activities. Classifications of levels of involvement are discussed in detail subsequently in this chapter. The information presented here was gathered from the first phase of data collection and a greater qualitative investigation in the second phase uncovered further details about the experience of leaders in using the internet. In describing their personal internet experience owner-managers pointed out that they were many times overwhelmed with information and did not know how to navigate through cyber space. Those with children highlighted that the experience of watching their children use the web inspired confidence about how to us it. The biggest restricting factor seemed to be uncertainty about the outcomes of certain activities on the internet and therefore many respondents had chosen to only engage in basic emailing and web searches. Most owner-managers attributed this to their family background and exposure which was primarily steeped in traditionalism. They therefore saw some internet activities such as purchasing personal items as being high-risk.

Risk-taking and the Owner-Manager

Risk perception emerged as being an important factor for the respondents. With most respondents perceiving internet buying and selling activities as risky, it is not surprising that they also perceive a relatively high risk in making an investment in platforms for carrying out e-commerce transactions for the firm. To further highlight this point, all of the respondents felt that an internet investment was either medium (45.2%) or high risk (54.8%). This suggests that there was some ambivalence or at least uncertainty about whether returns on such an investment were worth it. This refers to three types of perceived risk: psychological risk (Eastlick and Lotz, 1999) time risk (Tan, 1999) and financial risk. All three were applicable as owner-managers were apparently worried about the risk of losing time and money but also about the psychological risk of becoming too dependent on the internet and losing the "soft side" of their service provision. It is apparent that an exploration of how respondents view themselves as risk-takers in a competitive environment is critical. This finding was determined from asking respondents how they ranked themselves as overall risk-takers.

Table 4-5 Owner-manager's Risk-taking

Personal Ranking	Frequency	Percentage	Cumulative Percent
Very low	2	6.5	6.5
Low	21	67.7	74.2
Neutral	6	19.4	93.6
High	1	3.2	96.8
Very high	1	3.2	100.0
Total	31	100.0	

An overwhelming majority considered themselves to be low risk-takers with only 2 respondents ranking themselves as high risk-takers in *Table 4-5*. These "high risk-takers" represent those who were the previously mentioned avid internet users (see *Table 4-6* below).

Table 4-6: Cross-Tabulation: Personal Technology Use and Owner-Manager's Risk-Taking

Average Rate of Usage	Very Low Risk-Taker	Low Ris	Neutral	High Risk-taker	Very High-Taker	Total
Once per week (Emailing)	2 40%	3 60%	0 0%	0 0%	0 0%	5 100%
Twice per week (Emailing)	0 0%	10 83.3%	2 16.7%	0 0%	0 0%	12 100%
Every other day (Emailing and web browsing)	0 0%	7 70%	3 30%	0 0%	0 0%	10 100%
Every day (Emailing and web browsing)	0 0%	1 50%	1 50%	0 0%	0 0%	2 100%
Every day (emailing, web browsing and online purchasing)	0 0%	0 0%	0 0%	1 50%	1 50%	2 100%
Total	2	21	6	1	1	31

As seen in *Table 4-6* those owner-managers who viewed themselves as high risk-takers or very high risk-takers were using the internet extensively for their personal activities which included online purchasing. The converse was also true in that those who considered themselves to be very low risk-takers were in the category of the low users of the internet. While causation cannot be proven here this finding indicates that low internet users are also the ones who are less likely to engage in taking risks in investing in new technologies. This presents a serious challenge for the adoption of the internet for sales and marketing in these firms as owner-managers who describe themselves as low risk-takers also view investments in online platforms as being high-risk. So in addition to being low risk-takers in general they view technological investment as high risk. A relationship was observed between these risk perceptions and education levels.

Table 4-7: Cross-Tabulation: Highest Qualification and Internet Sales Investment Risk

Highest Qualification	High-Risk	Medium Risk	Low Risk	Total
Secondary	6 100%	0 0%	0%	6 100%
Diploma	11 100%	0 0%	0 0%	11 100%
Bachelor	0 0%	9 100%	0 0%	9 100%
Masters	0 0%	5 100%	0 0%	5 100%
Total	17	14	0	31

The findings in *Table 4-7* reiterate the earlier point that greater exposure levels through formal education processes have some influence on perceptions about the possible returns on investment (ROI) from engaging in online selling activities. In addition to education and technology experience, the nature of owner-managed businesses tends to facilitate greater levels of risk-aversion. An additional element which has been inferred through discussions with respondents is that they are unwilling to take risks with the business that directly affects the survival of their families. This is a due to the fact that these leaders are owner-managers of small, autonomous firms and there was a recurring emphasis on safeguarding "my business". Some respondents pointed out that they were likely to be less risk-averse if they were managers in a business which they did not own. It may be argued that they are not inherently low risk-takers due to their culture and society but rather act out of protection for a firm in which they have invested heavily. Alternatively they may be fearful about acting and investing in something for which they lack understanding.

Many of these protectionist behaviours result in a fixation on short-term benefits. For example one respondent articulated that "I have to make sure at all times that the business which my family depends on is safe so I cannot make hasty decisions." Having discovered these tendencies in phase 1 of the data collection process, it was necessary to identify leadership classifications and characteristics from the literature which may be applied in the context of technology adoption. The two dominant classifications which emerged were the transactional and transformational leadership types. The second phase of primary data collection therefore focused on identifying how these leadership concepts were applicable to technology adoption practices in these small firms.

Transactional versus Transformational Leadership

Transformational leaders uplift the morale, motivation, and morals of their followers while transactional leaders cater to their followers' immediate self-interests. According to Bass (1999) a considerable amount of empirical research has been completed since then, supporting the utility of the distinction between the two forms of leadership. He further articulated that changes in the marketplace and workforce over these decades have resulted in the need for leaders to become more transformational and less transactional if they were to remain effective. Leaders were encouraged to empower their followers by developing them into high involvement individuals and teams focused on quality, service, cost-effectiveness, and quantity of output.

This work identifies that new initiatives such as the adoption of innovations like technology are contingent on elements of what have been identified as leadership needs in the mainstream. For example the creation of "high-involvement individuals" and cost-effectiveness are required for the successful introduction, implementation and continued usage of technological tools.It must be elucidated here that the transformational leader is described as moving the follower beyond immediate self-interests through *idealized influence*, *inspiration*, *intellectual stimulation*, or *individualized consideration* (Bass, 1999). For the purpose of this study which essentially focuses on technology adoption, it was determined that the element from the four constructs identified above which is most critical is *intellectual stimulation*. This is displayed when the leader helps followers to become more innovative and creative. *Idealized influence*, *inspiration* and *individualized consideration* (one to one attention to followers),all address the relationship and kinship that followers (employees) attach to leaders. While these all result in followers aligning themselves to the goals of the leader, the only one which addresses a cognitive and behavioral change in the follower is *intellectual stimulation*. In order for employees to accept and understand innovative changes in the organization as well as participate in these changes, the leader must be able to stimulate creative thoughts and behaviours.

Although a dominant construct which applies to this research has been identified, all of the categories were explored in the second phase in order to enable bold statements about whether the leaders under investigation fit squarely into any of the two typologies of leadership. This concept addresses issues of admiration, respect and trust for the leader. In this research, there is an indirect relationship with levels of technology adoption within the firms as it indicates whether employees are likely to align themselves to the vision of the leader. This may relate to new innovations; however it is a more general issue of kinship and whether employees are for or against the leader, which will ultimately affect the desire to quickly buy into the vision for a new process or tool. Idealized influence is similar to charisma and followers tend to be more inclined to make change processes easier on a leader for whom there is great admiration. Owner-managers generally did not view the admiration of their employees as being important to the success of their businesses. They generally felt that employees were being paid to do their jobs and therefore it did not matter if they held them in high regard, as long as the job was done. Some respondents stated:

> "I am not here to be their friend. This is a business and I am more interested in employees doing what they are paid to do. I just want a profitable business and not one in which I am a role model."

> "I can understand that everyone has their own issues but I cannot be expected to be a counselor and mentor when I have to focus on making sure that we are still in business from day to day."

What emerged was the short-term orientation of the respondents. With the exception of two respondents, they were not particularly interested in the long-term acceptance of the vision but were more focused on day to day transactions and activities which employees were required to carry out. This is symptomatic of transactional leaders who focus on the task at hand and meeting the short-term objectives. On the other hand the respondents felt that while admiration was not vital, the respect of their employees was paramount. This ensures conformity to immediate processes and the following of instructions about what to do in the firms. A representative statement from one respondent was that “While you don’t have to like or admire me you must show me respect”. Any employee who shows disrespect is immediately terminated. A more sustainable approach may be to get followers to align themselves to the leader’s vision for the firm through not only coercive approaches but through admiration for the leader and leadership style. This would be particularly useful where major change will take place. For instance if firms were to engage in the introduction of a completely new platform for sales and marketing which completely eradicates the need for traditional approaches, it would be easier where there is a belief in the goals and aspirations of the leader.

The issue of trust was defined by respondents as employees’ belief in the fact that they are truthful and reliable. They thought this was very important because they wanted employees to feel that they “say what they mean”. This involves delivering on promises and also giving truthful accounts of an incident or event. Therefore if an employee is promised an incentive for carrying out a task they should have confidence in the leader to deliver on that promise. Very interestingly the use of incentives and rewards for carrying out specific tasks has been aligned to transactional leadership (Pawar, 2003). Transformational leadership appeals to intrinsic motivation that causes employees to see the overall vision and exceed minimum expectations. The three constructs needed to create idealized influence in firms do not receive equal attention from these owner-managers. While respect and trust are treated as critical elements of their leadership style, admiration is treated with very little regard as evidenced in the earlier quotations. This stems from a limited view of the meaning of admiration. Most respondents simply viewed it as whether employees “like me as a person”. Admiration goes beyond this and is closely aligned to the ability of the leader to set high standards and lead by example in accomplishing these standards. It is more than liking the person but rather being impressed with their abilities and actions. The data analysis reveals that the majority of these leaders do not rank very impressively with *idealized influence* which helps in creating an environment where new initiatives receive greater acceptance and involvement of the followers, who closely align themselves to a charismatic leader.

The construct of *inspiration* or *inspirational motivation* relates to how leaders are able to convey their expectations to employees in a way that gets them to understand and act towards meeting these expectations. It was applied in this research through an exploration of general approaches to communicating and achieving expectations as well as how these specifically relate to technology diffusion and adoption in these firms.

Respondents make their expectations of employees clear by conducting orientation sessions or having one to one meetings to explain expectations about specific tasks. They articulated however that they did not have sufficient time to ensure that there was complete understanding before tasks were carried out. An assessment is only done after tasks are completed. This many times results in trial and error situations which lead to time-wasting and frustration from employees about the processes. As stated by a respondent "sometimes because of the nature of our type of business we do not have a lot of time to spend doing non-transaction type activities." This suggests that there is very little time spent in preparation for any new initiative and is particularly important in attempting to understand the responses of employees to the introduction of new technologies.

The scenario represents one in which many of the owner-managers themselves are not enthusiastic about new technologies, and in addition do not take the time to communicate expectations effectively about their use. "Most times I cannot genuinely promote these new things because I am not impressed by them." Those leaders who introduced websites in their companies were asked how this process was carried out in terms of the communication of expectations, orientation and training. The dominant approach which was taken was one which involved an orientation session conducted by an external website developer after which owner-managers would encourage employees to become familiar with it on their own. In some cases rewards were offered for employees who were able to use the websites to stimulate sales. In an environment where these agents rely on commission, this was not a useful incentive as most of these sites were not equipped to handle online sales. The reality is that these leaders did not sufficiently stimulate employees to use the technologies because they were not necessarily attaching enough importance to the adoption process. In an optional situation the agents therefore made the decision to focus on what affected their salaries. Rogers (2003) denotes that types of innovation-decisions are optional, collective or authority driven. This generalization for example is quite useful as the observation of these service firms reveals that *optional innovation-decisions* very rarely result in comprehensive adoption for the social system. *Collective innovation-decisions* usually gain more traction but are largely leadership driven even though there is consensus by the team. This consensus usually stems from a charismatic leader (Hitt et al., 2001) or an opinion leader (Rogers, 2003).

The reality is that while *authority innovation-decisions* may yield compliance, those with power may also be opposed to the diffusion of a particular innovation. This again magnifies the importance of leadership in the innovation diffusion discourse. It is then fair to say that collective and authority innovation-decisions are more common in many organizations and that these processes are leader-initiated, which calls for an emphasis on the orientation of leaders and decision-makers. The disinterest which was displayed by leaders led to a general lack of enthusiasm for the employees and this explains why the majority of firms which had started websites reported that these were now inactive sites. Some respondents stated that they get employees to understand their strategic vision by placing a vision statement in a visible position of the office. They however did not follow-up to see if employees in fact understood what these statements were saying or if they shared in this vision. Most of these vision statements are also typically broad, vague statements and would not address specific initiatives such as the use of a new technological tool to better serve markets. In order for complete diffusion and adoption of technologies to take place, there needs to be greater focus on the effective communication of expectations, training of how to use them, and monitoring and evaluation of their use.

The firms in the study are small firms with the majority having less than ten (10) employees. This makes conditions favourable for individualized consideration or one-to-one attention to be given to each employee. According to Bass and Avolio (2003) this type of attention shows that the manager cares about the needs of the employee. This is referred to in the organizational decision-making literature as emotional intelligence (Mandell and Pherwani, 2003) and emphasizes that employees are likely to be more loyal to a manager if they get a sense that the manager sincerely cares and wants them to succeed. While this does not automatically translate into adoption behaviour, it sets a foundation upon which leaders may seek to engage employees in processes which may be new or innovative. This statement assumes that other conditions such as training, coaching and leading by example are also in place. Owner-managers indicated that they were too busy conducting the affairs of the business to give individual attention as they were integrally involved in operations. Individualized consideration takes a considerable amount of time and energy and involves knowing each employee, their abilities, their personal and professional goals, and what motivates them to perform at an optimal level. The respondents, while recognizing possible long-term benefits, acknowledged that they were only able to focus their efforts on activities that would bring relatively immediate benefits to the firm. "As I said before I have to concentrate my efforts on running the day to day business." Some managers also felt that it was a considerable waste of time to intimately know each employee when turnover rate was so high and employees would likely resign or be fired soon.

It is evident that the direction of the causal relationship may be blurred in the minds of the respondents. Perhaps the lack of connection with employees is a causal factor in the high employee turnover rate and greater attention to this may create longer lasting relationships between the firm and employees. Instead of viewing the turnover as the effect the respondents highlight it as the cause for their unwillingness to know their employees since relationships are typically short-lived. In an environment where leaders and followers are disconnected, the respondents pointed out that their approach to getting employees to perform well at their tasks was through authoritative coercion or incentives. One owner-manager stated "I think their salary should be their motivation and if they do not meet expectations then we cannot work together." Notably motivation drivers are being applied without an understanding of what drives each employee. If the wrong motivating factor is used then the approach is likely to be ineffective and the tasks may only be done at the minimum requirement level.

It is particularly striking that these approaches to encourage performance are taken as normal everyday activities for two reasons. One is that the use of incentives to encourage employees to carry out their normal duties is characteristic of a transactional leader, who simply focuses on a task at a specific point in time and does not act as a visionary. One implication is that if employees are being paid to carry out particular activities and then each task has to have an incentive appended then the firm will lose considerable revenues. Another implication is that employees are likely to perform below required standards where these incentives are absent. In addition to this type of incentive pointing to transactional leadership, there is a second, more far-reaching practical implication for these firms. The fact that the owner-managers have difficulties in encouraging good performance for regular, traditional duties makes it safe to say that an even greater challenge will emerge where new innovative ideas are introduced. This was demonstrated with the failed website attempts in firms, but the bigger issue is likely to be in future endeavours for moving firms higher up on the adoption hierarchy. Each stage of adoption is likely to yield greater resistance as it usually means that there will be a greater deviation from the norm. For example, a typical firm in this sample that only uses the internet for emailing marketing and promotion information to clients is likely to experience resistance at the use of websites for online marketing and sales, and even greater resistance to interactive social media marketing. This necessitates more creative ways of encouraging buy-in to the process from a leadership perspective, and the narrowing of the distance between the manager and employees through individualized consideration may provide a platform from which catalytic changes many take place in how employees view work in the organization and boost performance levels to maximize potentials.

The existing conditions of manager-employee relationship are symptomatic of transactional leadership at play. This work takes a hierarchical approach with the two forms of leadership rather than accepting the view that these types of leadership are at extreme ends of a continuum. This means that transactional leadership traits provide important preconditions for building transformational leadership skills. This distinction is important because the dominant transactional traits which now exist in these firms, such as meeting daily targets and controlling the work environment do not run counter to the transformational paradigm but rather provide a basis upon which this type of leadership may take the firm to the next level of technology adoption. Having made this distinction, it is critical to point out that these transactional leadership traits are simply not sufficient to encourage greater participation in innovative activities as they are primarily focused on maintaining the status quo within organizations. To present a prescriptive approach to encouraging greater participation and involvement of employees makes the assumption that the managers themselves want change to occur. In some cases owner-managers were not embarking on a path to change but were rather trying to survive using antiquated approaches in the hope that the industry and markets will revert to what they used to be. In these cases there was no attempt at intellectual stimulation for new and innovative ideas from within the organizations.

Intellectual Stimulation

The construct from the Multifactor Leadership Questionnaire which most applies to this research is intellectual stimulation. This is displayed when the leader helps followers to become more innovative and creative. An exploration of the other three constructs was important to be able to clearly identify if these leaders would fit into any of these two leadership categories based on all four constructs. While the previous three demonstrate the likelihood of followers aligning themselves to the goals of the leader, the only one which addresses a cognitive and behavioral change in the follower is *intellectual stimulation*. In order for employees to accept and understand innovative changes in the organization as well as participate in these changes, the leader must be able to stimulate creative thoughts and behaviours. The current practices of the firms are steeped in traditional practices which are not producing the desired results and there is a need for new ideas to emerge. One area which is stagnant and needs to become more dynamic is the use of a singular static platform for reaching customers. A move towards ideas surrounding online platforms and internet technologies which may improve the reach and appeal of firms must be influenced from the top of the organizations.

If the owner-managers themselves lack innovative ideas, they must create an atmosphere where creative thought processes are fostered and allowed to flourish. Additionally a lack of ideas from the managers does not mean they cannot influence creativity and innovative behaviours through their leadership style and initiatives that facilitate cutting-edge ideas. In response to how employees are encouraged to generate new ideas or come up with new approaches to job tasks, the managers overwhelmingly stated that this was not encouraged. It was articulated that these ideas only came from the top of the organization. The organizational culture of the firms promoted absolutely no deviation from standard operating procedures or even suggestions of how a task or process may be improved. Respondents stated:

> "While getting ideas from employees may be a good thing, it is more important to me that they are not distracted from doing what they need to do which would be sales. In any case if they suggest something based on their like or dislike for technological innovations that would not be a priority issue."

> "Maybe if someone has brilliant new ideas for introducing more innovative things then they are probably over-qualified for this job and would be a better fit somewhere else. I just want them to get on with the business of meeting customers' needs."

Innovative solutions to problems in the firms were either non-existent or very minimal due to the dependence on owner-managers, who were traditionalist thinkers to create contemporary, cutting-edge solutions. A more open system of idea-sharing would allow for a greater number of options from which to choose and would also empower employees to feel like they are making a valuable contribution to the development of the firm. This would ultimately make the adoption of these ideas easier to "sell" to employees if they feel like a part of the process and if an innovative strategy was generated by a peer. With the exception of two respondents it is fair to say that according to the major leadership constructs which were employed, owner-managers ranked low in all the areas and can be generally categorized as transactional leaders who simply seek to monitor and control. The interviews about leadership and technology adoption also revealed that two generic categories are not sufficient to classify leaders in the innovation and adoption of technology discourse.

While the constructs from the Multifactor Leadership Questionnaire were useful in identifying general transformational or transactional leadership traits, it was felt from looking at other measurement instruments such as the work of Singh and Krishnan (2007) who attempted to develop and validate a new scale on transformational leadership that adaptability is critical. While they used the category of "Conviction in Self" to highlight the ability to plan in advance for the worst possible outcomes, this work recognizes that this stems from the ability to adapt and create strategies that are flexible. Respondents stated that they did not necessarily have a system for determining the effects of their current decisions in the medium to long term. Having been in the industry for many years a recurring theme was that of intuition. Owner-managers felt that their decision-making is driven by what they call "gut-feeling" based on experience with how the industry operates. A representative response from one respondent highlighted that "I have been in the business for a long time and most times I just know if something feels right. If it feels right and I can afford it then maybe." This signifies a reluctance to think strategically and it was not surprising that they also stated that they did not have a way of preparing for the worst possible outcomes, but rather dealt with a crisis when it happened. This reactive approach saw massive decline in the number of businesses as well as in the profit margins of the remaining ones when their major source of income was cut by airlines.

Contingency planning is therefore lacking in these firms and their reaction is sometimes slow when significant change takes place. One of the changes to which they have reacted fairly slowly is the technology revolution and more specifically the internet explosion. This slow reactive approach was also seen in the fact that when respondents were asked how they identify when old approaches no longer work and new ones are needed, they stated that sometimes there was no way to tell until it was too late. Once again the lack of a clear system for evaluating the needs of the firm makes it particularly challenging for the formulation of agile strategies, which allow for fundamental change to take place. The assumption is often made that technology will provide the solutions needed to help firms to survive. It may be argued that owner-managers may recognize that some types of technology are not suited for their firms and therefore make the choice not to expand into those platforms. Such a determination would need to be made based on solid data from needs assessment exercises, rather than what currently exists where it depends on an owner-manager's personal preferences. The evidence resoundingly points to a need for something new to happen in light of previously mentioned conditions under which these firms operate.

These agencies, which have seen a decrease of more than 50% in the number of companies in operation over the last ten years, typically cater to the outbound market. Their ability to leverage their importance may however come from an improved capacity to meet the huge need for an online presence for the country. At present these agencies receive little attention from statutory bodies and a greater contribution to the island's tourism may present a strong case for governmental collaboration and support. In such a volatile marketplace, there is a greater need for leaders of these firms to become visionaries rather than just maintain their current position as transactional "caretakers" of static procedures.

Leadership and Strategy

There has been an abundance of research which connects strategy to organizational success. For example in his seminal work Porter (1985; 2001) points to five competitive forces, which have been extensively used by researchers (David, 2007; Kim et al. 2004; Stonehouse and Snowdon, 2007; Poon, 1993; Wynne et al. 2001) over the last three decades. Much of the research however has indicated that firm strategy must be influenced by competitors as well as markets. While these are useful arguments they are both output driven discussions with very little emphasis on inputs such as ownership, leadership and resources. It is critical that this work fills the gap in discussing input considerations. Fundamentally the strategy is essentially influenced by the markets and the competition, however there needs to be more emphasis on how leaders respond to intelligence about these two factors. There were respondents who recognized a market trend and acted differently from others who were aware of the same trend, which makes an investigation of leadership responses to this kind of information very important. Why for example, were some owner-managers content to accept that the local markets were not ready for online platforms and in response did not see the need to diversify distribution; while others with the same information, felt that they should be proactive in anticipation of market changes?

Of particular interest was the fact that most managers felt that there is a high level of incompatibility with having a high-tech strategy coupled with a hi-touch one (intense personal interaction) despite compelling arguments by theorists that a multi-platform approach to distribution will yield sound business models. The owner-managers in the sample seemed more concerned with what Eastlick and Lotz, (1999) refer to as a perceived psychological risk of becoming too dependent on the internet and losing the essence of their customer interaction. As a respondent stated: "Too much technology can take away from the personal touch that we normally provide. I would hate for my company to become so reliant on these machines that we forget the importance of people in all of this."

It is clear therefore that if these owner-managers are to engage in greater internet technology adoption, they will need some form of assurance that online and offline channels of distribution can "peacefully co-exist". The reality is that since these owner-managers highlight that they typically make decisions unilaterally, there would need to be a change in the leadership of the organizations. This change may come through a shift in leadership style which may be fuelled by education or greater exposure, as will be discussed later in this chapter.

A recurring theme in the interviews was the limitation placed on these small firms due to a lack of sufficient resources. The actions of the leader become important in decision-making regarding resource allocation (Kraajenbrink et al. 2010). The findings are in support of this work and this is even more magnified where firms have significantly limited resources as tough decisions must be made about priority spending in order to efficiently use these resources. The resource limitations which were highlighted by owner-managers focused primarily on human and financial resources. In most cases it was felt that there was not enough money to invest in a hi-tech strategy. In other cases however managers felt that where money was available they did not have capable human resources to consistently maintain technological innovations such as websites. This was also evidenced where firms in the minority which had websites stated that these were inactive. One respondent point out that: "Even if I wanted to introduce these high-tech approaches the level of capabilities of my employees would not be able to support these activities. The entire thing would be too much of a big investment." The respondents highlighted that with the limited financial and human resources they were more inclined to use resources in areas that would assist in the day to day operations rather than on strategic initiatives which they were not sure would yield significant returns. Technological innovations were mentioned as being risky investments which were more geared towards the long-term and many respondents argued that it was not considered a priority area since it did not address the immediate needs of the markets being served. This highlights a transactional approach to leadership, which simply seeks to enforce procedures and control daily activities.

The situation for the firms under investigation is also a unique one in that the respondents are owner-managers who have a stake in the immediate or short-term benefits of the operations. This work posits that owner-managers in a stand-alone environment are inherently transactional leaders based on the challenges of the industry coupled with their desire to realize immediate returns on investment as the money feeds directly into family survival. This is an extension of the entrepreneurship literature and particularly the work ofCovin and Slevin, (1988) which argues that small business managers prefer to be decidedly risk-averse, non-innovative, and passive or reactive. Carland et al, (1988) state that they operate their business as an extension of their individual personality and immediate needs. In addition to a reluctance to adopt internet technologies, this is evidenced by the sample's refusal to engage in inbound travel due to a lack of commission although this may yield long term benefits.

The exceptions to this norm were companies that had numerous branches and were established brands or companies which were affiliated with other business in other industries. It appears then that the process of prioritization for resource allocation is influenced by owner-manager's personal needs, preferences and short-term fatalistic goals, which do not consider a strategic vision for these firms. Hence, it is important to create a profile of leaders which identifies characteristics that can enable movement to higher levels of technology adoption for sales and marketing in particular. This is critical in an information-intensive travel industry, particularly in owner-manager, small firm, developing country situations where decision-making is so heavily dependent on the likes and dislikes of the individual and ultimately the leadership style employed. The following section elaborates on the characteristics identified.

Leadership Characteristics for Technology Adoption (Owner-Managed, Small Firms)

A number of variables have been identified in this study as well as in previous research (Burns, 1978; Karnes and Chauvin, 1985; Bass, 1999; Bass and Avolio, 2003; Singh and Krishnan, 2007) about factors that apply to the major leadership categories. Very little emphasis has been placed on identifying leadership characteristics at various levels of technology adoption. For example the work of Peterson et al. (2009) has identified that some transformational leadership traits which apply to CEOs who lead high-technology start-up firms, but until now research which has challenged these broad leadership typologies along a hierarchy of technology adoption is absent. This kind of detailed analysis of a leader at each adoption level, which is found in this book, will be instructive about what characteristic is needed to move to the next step on the ladder.

In the first instance leadership emerged as the dominant driver for these owner-managed small firms for a number of reasons. A comprehensive approach was taken through the lens of *Critical Social Science*which allowed for an exploration of a number of micro and macro factors that may affect adoption levels within societies and firms. These included culture, the digital divide, strategy and resources. National culture emerged as a peripheral contributor since it did not explain differences in the behaviour and performance of the owner-managers. The digital divide also did not emerge as a central driver given that the Jamaican society has a high level of mobile penetration and in particular smart phone use with equal access within the society for businesses. There is still a gap in the creation of innovation however the access gap has been greatly reduced. The previously existing access gap related more to the use of computers. The divide however plays a role at the level of the destination management organization in the context of tourism where there is still a need for a more high-tech approach to meet the information and booking needs of potential high-tech tourists.

Those broader macro factors which did not emerge as significant would have been the most influential in explaining constraints on the leader. The other micro factors of strategy and resources are however determined in large part by decisions made in the firms. In the case of these firms, owner-management and their size result in simple decision-making processes which are driven by these leaders in the organizations. The strategy to be employed and the allocation of resources as explained earlier in this chapter are driven by the priorities of these leadership figures. In light of the pivotal role of leadership, some key characteristics were identified at each stage of technology adoption within these small, owner-managed, travel firms.

Education

An important factor from the interviews was education level and type. Those owner-managers who were inclined to adopt based on immediate plans as well as a general intention to explore other commercial uses of the internet, tended to have university qualification at least at the bachelor degree level. Their openness to a greater use of internet technology in sales and marketing comes from exposure to the use of technology in a university environment as well as formal discussions surrounding the technology revolution and its potential benefits.

While some of these respondents were not using cutting-edge technology at the moment, they were not resistant to the notion that it has a place in the firm. This presents a particularly fertile condition for the birth of new processes in these firms over which they have control, such as website use for e-commerce and social media for sales and marketing.

The type of education also played a role in the openness to technology adoption. Interestingly it was found that those owner-managers who had generic business degrees were much keener on utilizing multiple marketing and distribution platforms than those with specific travel and tourism qualification. This may speak to a greater emphasis on competitive strategies and the development of sound business models in business studies. This then raises the question of whether tourism degrees suffer from a tunnel-vision which does not embrace a more fundamental understanding of key business concepts. This presents an area for future research.

Previous Work Experience

The professional background of the respondents seemed to differ greatly between likely adopters and unlikely adopters. The two most technologically progressive thinkers in the sample had worked outside of the travel industry extensively. The market leader in the sample had extensive experience in real estate and had built and operated a successful company in that industry for many years. The other respondent who had not yet implemented many of the plans but had intentions to do so was a bank manager for many years who used retirement funds to start a travel agency.

On the contrary, the slow or low adopters had spent their entire careers in the travel business. Some started out as entrepreneurs while others worked for other travel companies and airlines before starting their own operations.

This fact may be due to a reason similar to the education variable, where a greater exposure to business practices which were thriving in other industries created transferable approaches to competitiveness. The real estate and banking industries experienced an evolution and restructuring, which may have taught these owner-managers some valid lessons. It must be noted that this work does not posit that all owner-managers should immediately employ internet sales and marketing approaches. However it advocates that there must at least be a certain level of openness to ideas which may enhance the relevance of their firms especially at a time where this is being questioned. The approach of many struggling agencies in this study seems to be "wait and see".

Technology Experience

It was found that the personal experience of these owner-managers within firms with using the internet provided strong insights into their propensity to introduce it as a sales and marketing tool in their firms. The rate of use for owner-managers in the sample was quite low as demonstrated in *Table 4-4*. The rate of use does not say much on its own though and an exploration of the reasons for which it is used adds more robustness to the discussion.

The reasons for personal internet use for leaders were very telling as these translated into business practices and firm adoption behaviour. The only firm which had engaged in online sales was one of the two firms which had internet savvy leaders who were engaging in online purchases of their own. The other was in the planning phase for platform diversification. It must be noted however that their personal Facebook use did not translate into firm use as social media was still viewed as just a "social" tool. On the other hand, the firms whose leaders engage in minimal internet use for emailing up to two times per week experienced low levels of internet involvement and primarily used the web for email activities.

The personal internet experience of the latter group of owner-managers was filled with overwhelming information, which made it particularly challenging for them to navigate through cyber space in an effective way that could enhance decision-making. This resulted in an overall resistance to a phenomenon which they did not completely understand. The final outcome was that low personal technology experience led to similarly low usage in these firms which were essentially an extension of the owner-managers.

Risk Aversion

The perception of risk emerged from the data as being an important factor for the respondents. An overwhelming majority considered themselves to be low risk-takers with only 2 respondents ranking themselves as high risk-takers. These "high risk-takers" represent those who were the previously mentioned avid internet users. This presents a serious challenge for the adoption of the internet for sales and marketing in these firms as owner-managers who describe themselves as low risk-takers also view investments in online platforms as being high-risk.

An additional element which has been inferred through discussions with respondents is that they are unwilling to take risks with the business that directly affects the survival of their families. This is a feature of the fact that these leaders are owner-managers of small, autonomous firms. They were more deeply involved from an ownership perspective and pointed out that they were likely to be less risk-averse if they were managers in a business which they did not own. It may be argued that they are not inherently low risk-takers due to their culture and society but rather act out of protection for a firm in which directly affected their families.

Family Composition

The issue of the make-up of the family emerged from simply asking about the respondent's backgrounds. Inferences revealed that the leaders which were more interested in greater technology use were those with teenage children in their households. They highlighted that the experience of watching their children use the web inspired confidence about how to use it. The biggest restricting factor seemed to be uncertainty about the outcomes of certain activities on the internet and therefore many respondents had chosen to only engage in basic emailing and web searches.

This prompted further exploration to identify if there were differences with those who seemed less interested in increased adoption. The findings revealed that with the exception of one, all of the other low-adopters had older children who were now adults and did not live with them or had small children, who apparently did not influence them in this way. It would therefore appear that such a situational factor has a greater impact on the cognitive disposition of these leaders towards technology adoption rather than the broader issue of national culture.

All of the afore-mentioned variables in some way relate to situational considerations that generate differences in each individual. One major difference which has been identified is the level of intellectual stimulation which is provided by leaders of low-adoption firms and high-adoption firms; as defined by those who intend to adopt more cutting-edge tools in their businesses.

Intellectually Stimulating Traits

While all of the respondents ranked low on three of the transformational leadership constructs provided by Bass and Avolio (2003) in the MLQ, there was noticeable difference in the *intellectual stimulation* construct for owner-managers who had a greater inclination towards technology adoption. *Idealized influence*, *inspirational motivation* and *individualized consideration* are all related to creating an atmosphere of admiration, respect and kinship to the leader. Although there was no difference between respondents in these areas, the *intellectual stimulation* construct showed that the more innovative-minded owner-managers were different and ranked higher in this area.

This is displayed when the leader helps followers to become more innovative and creative. An exploration of the other three constructs was important to be able to clearly identify if these leaders would fit into any of these two leadership categories based on all four constructs. While the previous three demonstrate the likelihood of followers aligning themselves to the goals of the leader, the only one which addresses a cognitive and behavioral change in the follower is *intellectual stimulation*. The respondents that had intentions of improving online platforms for sales and marketing created an atmosphere in which employees were free to make innovative suggestions. This environment is conducive to the generation of a multiplicity of ideas which may help in the formulation of effective strategies.

Ideally owner-managers should endeavour to improve in all of these transformational leadership categories; however it is evident from this research that the *intellectual stimulation* element is sufficient along with the other variables identified by this investigation, to stimulate higher levels of technology adoption in owner-managed, small travel firms. This finding points to another more over-arching issue in leadership and technology adoption research. There is now the need for more specialized leadership typologies in relation to technology adoption behaviours, as the current broad classifications of transactional and transformational leadership do not necessarily apply directly in the innovation diffusion and adoption discourse. The following leadership classifications have been developed by the author to help in identifying the leadership characteristics and needs at various stages of technology adoption for small travel firms. These have been influenced by the transformational leadership constructs and the organizational decision-making literature which describes the decision-maker as creator, actor or carrier. Since this research has established that the key and sometimes sole decision-maker in owner-managed small firms is synonymous with the term leader, the two have influenced these classifications. This research however promotes a greater application as it relates to technology adoption for owner-manager operated small travel firms.

The following discussion proposes a synthesis of the findings of this research which allow for the profiling of leaders of owner-managed small firms at each stage of technology adoption. This is based on previously discussed characteristics in this chapter. The originality of this study emanates from some key gaps in the approaches to staged adoption studies. The first is that the body of research on stages of adoption takes a sequential continuum approach to understanding singular innovations. Secondly the adoption classifications from previous research tend to categorize firms in general. While this is useful, it does very little to account for small firms in which there is a great level of autonomy such as in the case of owner-managers. The third issue is that although there has been some research which has recognized the influence of leadership traits on small businesses (see for example Thong and Yap, 1995; Peterson et al. 2009), this research failed to identify leadership characteristics as a driver of each stage of adoption.

The work of Thong and Yap (1995) highlighted three important characteristics of innovativeness, IT attitude and IT knowledge. This however addressed overall IT adoption as opposed to a staged approach and this was at a time when less pervasive business platforms existed. Peterson et al. (2009) in applying transformational leadership traits to high-technology environments focused on generic traits in start-up firms. This research through an emphasis on owner-managed small firms advances the debate on staged technology adoption and the leadership typologies that influence each stage. Each component of the model being constructed will be explained in detail. At the end of this discussion a coherent, composite model will emerge in *Fig 5-7*. As highlighted earlier in this chapter some key leadership typologies emerged as it related to staged technology adoption in small owner-managed firms.

Leadership has been identified as the key determinant of technology adoption for small, owner-managed firms. Some key variables have been identified as being influential in technology adoption from a leadership perspective. This necessitated a new conceptualization for leadership typologies in the innovation diffusion and adoption discourse, as previous generic typologies failed to take some of these key variables into account. Additionally these newly developed typologies merged the leadership and organizational decision-making discourses to identify applicable, disaggregated constructs that will inform the model which will emerge in chapter 5. In the subsequent chapter the model will identify the various stages along a technology adoption hierarchy where these new leadership typologies may be applied. Each typology, namely **resistors**, **caretakers**, **stabilizers**, **reactors** and **transformers**, will correspond to technology adoption stages in firms and a detailed discussion of how the model was developed will ensue. Prior to this though, the overall conceptual development of the research will be explained.

CHAPTER FIVE

STAGED TECHNOLOGY ADOPTION: PRACTICAL IMPLICATIONS OF AN INNOVATIVE THEORETICAL MODEL

The conceptual framework presented in *Fig 1-2* has been amended to reflect a difference in the feedback loop for firms. Initially when overall firm adoption decision-making was considered, feedback information was viewed as a function of a system, which was designed to identify firm performance and make decisions in a systematic, consultative way based on strategy. The new conceptualization identifies that in small, owner-managed firms the feedback goes directly to a unilateral decision-maker, whose attitudes and personal beliefs determine technology adoption, as opposed to an objective evaluative system. This emerged from the finding that regardless of how a firm performs, the decisions about technology use are driven by personal preferences of an autonomous owner-manager. This paints the overall picture and is further discussed in this chapter, where the development of the model in *Fig 5-7* is addressed.

Theory and Concepts

Conceptualizations of technology adoption have resulted in major theoretical developments, which for the purposes of this research have been categorized as pre-internet phase and post-internet phase research. The pre-internet phase saw major contributions from Rogers (1962; 1976; 1983; 1994; and 2003) who developed on the work of Ryan and Gross (1943). Rogers' conceptualization involved four key elements of diffusion as (1) innovation (2) communication channels (3) time and (4) social system. This provided a platform for the growth of research in the area as it developed factors which were more applicable across fields.

Fig. 5-1: Revised Technology Decision-Making Input Framework

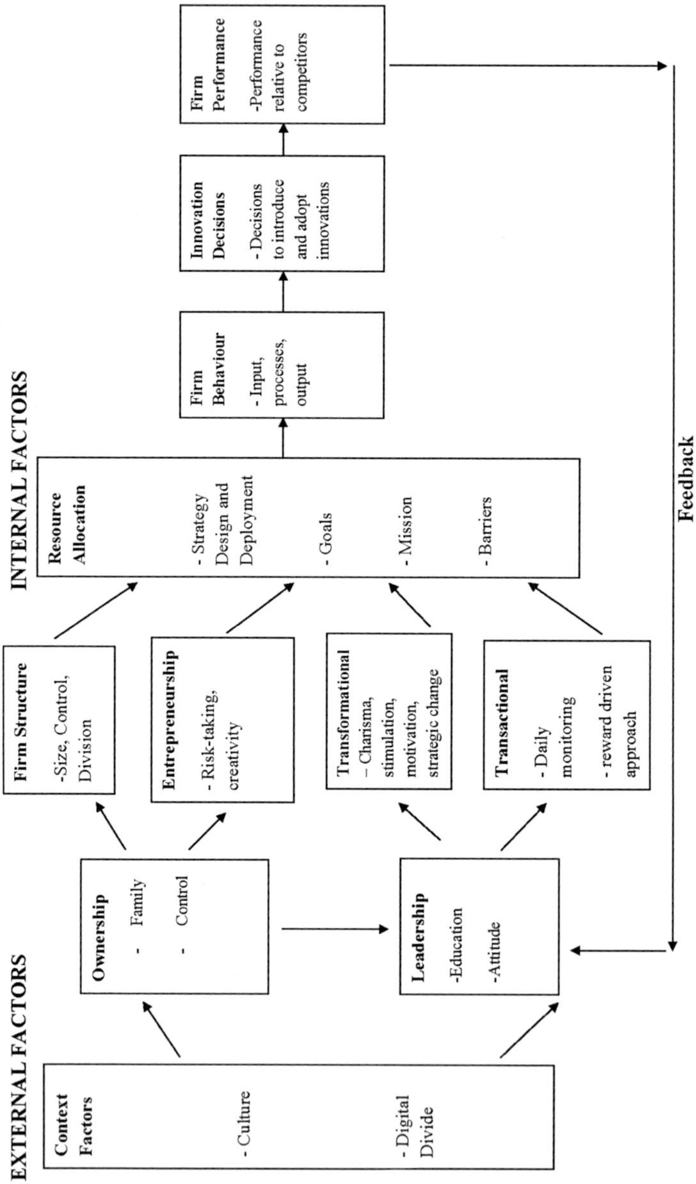

Much of the subsequent research posited that technology adoption was primarily positive (see for example Bagozzi et al. 1992; Lederer et al. 1998) and only fairly recently did Latzer (2009) identify that some technology innovations may be unsustainable or even disruptive. One of the most resonant works to emerge in the pre-internet phase after Rogers' seminal work was research by Davis (1989). In his technology acceptance model he argues that situational as well as personal influences such as perceived ease of use and perceived usefulness are critical in determining the level of technology acceptance. This added the elements of observation and perception to the discussion. It was the post-internet era however, which sparked most of the technology adoption debate across industries.

Much of this discourse has emphasized a transformation from an industrial economy to an information economy, and this may be seen in the development of the literature. Since the internet explosion in 1991, the research has been assessing the pervasiveness of the World Wide Web in a variety of industries. There is consensus by many theorists (Rayport 1995; Choi and Stahl 1997; Bakos 1998; Williamson and Scott 1999; Afuah and Tucci 2001; Applegate 2001; Wirtz 2001; Rappa 2002) that the rules of competition for established business have been transformed by the internet. Out of this emerged a focus on information asymmetry reduction, disintermediation and reintermediation. The latter two in particular received significant attention in travel and tourism (see for example Laws, 2001; Buhalis and Licata, 2002). This book found that internet transformation was fairly slow in the travel firms being studied and the technology adoption drivers and barriers are discussed in detail in this chapter.

The drivers of technology adoption have been varied in the extant literature. The conceptualizations have emphasized drivers such as national culture (Westwood and Low, 2003), the global digital divide (Stump et al, 2008; Minghetti and Buhalis, 2010), resources, (Brown et al, 2007), strategy (Stonehouse and Snowdon, 2007) and leadership (Elenkov and Manev, 2005; Lynskey, 2004; Peterson et al, 2009). All of these elements were accounted for and addressed in the secondary data collection as well as phase 1 of primary data collection. These factors were conceptualized in two coherent conceptual frameworks (see *Fig. 1-1* and *Fig. 1-2*). *Fig. 1-1* identifies fragmented relationships which exist in the literature which have typically focused on singular relationships. The findings of this research will now provide for greater coherence in conceptualizing technology adoption drivers for small, owner-managed travel firms.

There have been some gaps in addressing technology adoption in firms due to this fragmentation in previous works. *Fig. 1-2* illustrates a coherent framework, which places technology adoption decisions within the broader theory of decision-making and highlights relationships which may be explored for future research. This research further emphasizes the leadership role, as the focus of the study is on owner-managers of small travel firms, who have a great deal of autonomy in decision-making. It must be noted that while *Fig. 1-2* influenced the primary data collection process, some of the relationships identified were subsumed in other factors. Ownership was subsumed by leadership for example as the research focuses on owner-managers. Fundamentally the research process highlighted that an understanding of adoption drivers in firms lie in the demystification of the decision-making process. The question then emerged: which of these factors influences decision-making most significantly in firms?

The research has led to the creation of leader profiles and ultimately new leadership typologies which have been developed based on the generic transactional and transformational typologies. The new typologies, apply more specifically to owner-managers of small firms in information intensive industries. Each typology will be further explained in this chapter. Each typology relates to specific levels of adoption along an adoption hierarchy, also referred to as adoption stages. Stages in this research will be more vertical in nature as it addresses movement to higher levels of technology adoption.

Adoption Stages

The leadership typologies being advanced relate directly to particular stages of the technology adoption process. A considerable body of work exists regarding various stages of adoption in particular for online platforms (see *appendix A*). This research will identify stages through which a firm progresses for a particular innovation as well as through different levels of innovation adoption. This is different from previous research as it takes a multi-level approach that assesses the driver at each level. For example Rogers (1983) highlighted that a firm goes through stages such as: agenda setting, matching, redefining, clarifying and routinizing for each new innovation. Cooper and Zmud (1990) advocated a six-stage process to include initiation, adoption, adaptation, acceptance, routinization and infusion. Damanpour (1991) succinctly collapsed these stages into two key areas known as initiation and implementation.

Even the more contemporary research focuses on the stages of a specific innovation adoption. In attempting to advance a staged model, Daniel et al. (2002) focused on e-commerce adoption specifically and created firm clusters in a sequential manner. Cluster one to four involved developers, communicators, web presence and transactors. All of these represent different degrees of e-commerce readiness and adoption through the key variable of technology involvement. Even more recently Aquila-Obra and Padilla-Melendez (2006) have articulated a four cluster approach to internet technology adoption which also emphasizes a sequential process for a single innovation. For a more comprehensive list of adoption models please see *Appendix A: Summary of online adoption models.*

For the purposes of this research there was a need to conceptualize some technology adoption stages which represented a more holistic demonstration of technology adoption within firms. As opposed to previous approaches this conceptualization does not only consider single innovations and accompanying sequential processes but uses a hierarchical approach to highlight different levels of adoption with each level representing a more cutting-edge type of adoption. This was based on the findings of the research for the firms in the sample. While this is not the main focus of this study, it was necessary to develop a hierarchy that corresponds to the new leadership typologies which are ultimately being advanced. Each stage of adoption refers to back office use and ultimately sales and marketing use at that level. Subsequently a discussion of the leadership profiles will be discussed in relation to these adoption levels. The relationships will be demonstrated more closely in *Fig. 5-7*.

The Adopters

Given that this research is more concerned with vertical movement along an adoption hierarchy rather than horizontal movement along a sequential process, simple phases of initiation and implementation are adopted at each level of adoption. Initiation refers to the point at which firms consider the need for an innovation, search for information and explore resource allocation needs. Implementation addresses first use of the innovation and the point at which processes evolve to match the new adoption. The broad category of *computer adopters* was used to describe firms at the bottom of the hierarchy that were simply engaged in the use of computer terminals and hardware for back office accounting functions or for front office functions such as sales. Firms which do not use online sales tools other than GDSs are also placed in this category since their adoption was simply based on the free provision of the system by the supplier. It must be noted that computers are also an important pre-condition in firms for other innovation adoptions such as the internet as discussed in innovation interdependence in chapter 1. All companies had passed through the initiation and implementation phase at this level.

The next group on the adoption hierarchy is termed *internet adopters*. This group represents firms in which the internet is used for emailing and web browsing. A fairly large number of the firms in the sample had not moved beyond this level and only engaged with the internet for client communication and information searches. Most firms already went through implementation of this phase. Surprisingly there were still those which were at the point of initiation. *Website adopters* refer to the firms which have created and used company websites for general and marketing information sharing. Websites in this case typically provide static information. While a few firms in the sample had implemented these, they were eventually abandoned and are now inactive. Much of the information is now outdated and there is little intention to revitalize these efforts. Outside of those who already implemented the use of these websites, a few other firms were at the point of initiation while the majority of firms were resistant to its use. The next level of the hierarchy is called *e-commerce adopters* and refers to firms which use websites for actual bookings and payment. For this research, this level is considered to be key as it is the level at which there is strident refusal to adopt. The key investigation surrounds why firms are resistant to the notion of online selling. One firm is at the point of implementation while one other firm is at the point of initiation for this level of adoption. An overwhelming majority of firms do not aspire to this level and have no intentions of even exploring the option.

Reasons given for this in the findings include limited technology experience, limited knowledge of benefits and the perception that the need does not exist. Some responses included:

> "I would be wasting my time to try to sell online to my clients because most of them do not like using the internet to do serious business; furthermore they like having that personal contact and they wouldn't trade that for anything."

> "I know that there are people who buy online. I have never done so myself and I am kind of wary of doing business with machines. I think my customers also look at it that way and they prefer to talk to someone."

The views of the respondents represent limited approaches to strategy, which do not allow for changes outside of normal activities. The emphasis is on short-termed approaches to keep business alive. This is characteristic of the manager being the owner of a small business in which there is autonomy to make unilateral decisions. The result is the emergence of protectionist tendencies which involve low levels of risk-taking with a business that meets the immediate needs of these leader's families and interests.

The final level on the hierarchy has been identified as social media adoption in firms. *Social media adopters* would refer to firms in which social media is used for promotion and interaction which results in sales activities. While none of the firms in the sample fall squarely into this category, it appears likely that the firm which is now at the implementation phase of e-commerce adoption will move to the initiation phase of social media adoption. This is based on an expressed intention to explore such an option. It is important to note that social media is also being used on a personal level by this owner-manager. He states:

> "I currently use Facebook to connect with friends and I am not using it in the business as yet because I need to find out more about whether customers view it as a serious business tool or just as something to do for fun. I am open to the idea though and will look into it."

Most firms are resistant to moving to any adoption level which is higher than basic internet adoption in the form of emailing and browsing. There are a few firms which have adopted websites for general information and an even fewer number which are exploring online selling. The ultimate aspirational level of social media adoption is still in its infancy and only one firm is even exploring the option. The differences between firms at each level will be highlighted and correlated with leadership profiles.

Contribution to Theories of Staged Technology Adoption

While these categories will assist in the discussion of the hierarchical movement of technology adoption in firms, they are not the essence of what is seminal about this work. The originality of this study emanates from some key gaps in the approaches to staged adoption studies. The first is that the body of research on stages of adoption takes a sequential continuum approach to understanding singular innovations. Secondly the adoption classifications from previous research tend to categorize firms in general. While this is useful, it does very little to account for small firms in which there is a great level of autonomy such as in the case of owner-managers. The third issue is that although there has been some research which has recognized the influence of leadership traits on small businesses (see for example Thong and Yap, 1995; Peterson et al. 2009), the body of research research failed to identify leadership characteristics as a driver of each stage of adoption. These will be discussed in the subsequent section.

The work of Thong and Yap (1995) highlighted three important characteristics of innovativeness, IT attitude and IT knowledge. This however addressed overall IT adoption as opposed to a staged approach and this was at a time when less pervasive business platforms existed. Peterson et al. (2009) in applying transformational leadership traits to high-technology environments focused on generic traits in start-up firms. This research through an emphasis on owner-managed small firms advances the debate on staged technology adoption and the leadership typologies that influence each stage. Each component of the model being constructed will be explained in detail. At the end of this discussion a coherent, composite model will emerge. Some key leadership typologies emerged as it related to staged technology adoption in small owner-managed firms.

Resistors

Fig. 5-2: Resistors

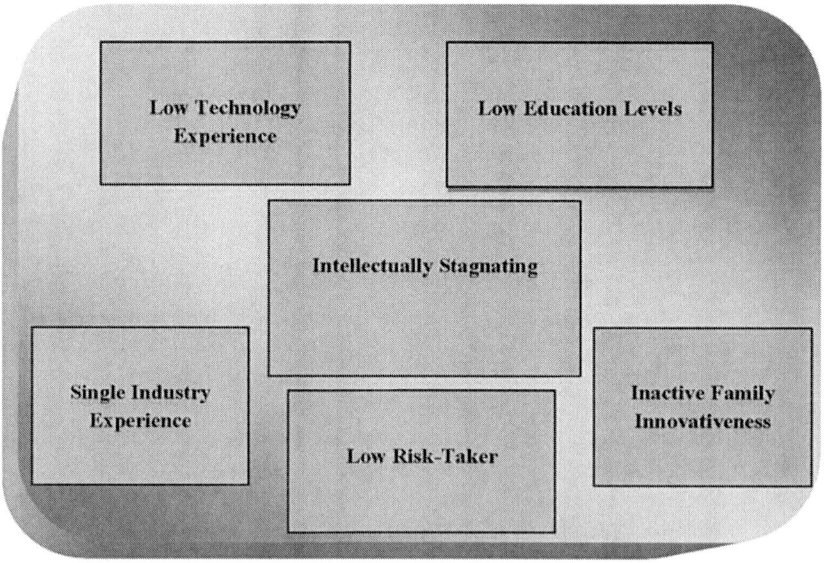

Research on the transactional leader has broadly pointed to a short term orientation in completing tasks at hand by using rewards to motivate employees to complete tasks. This research however highlighted that within this category there are different types of transactional leaders and some clear variables which separate them. At the lowest level is the **resistor** who is least likely to effect change in the firm and is more interested in maintaining traditional approaches. These persons rank themselves as low-risk takers as measured by openness to new ideas from employees, and say that they are willing to do only what is necessary to meet predetermined objectives. They have low technology experience which is evidenced by their lack of use of computer related technology with the exception of infrequent personal emailing and all of their work experience has been in the travel industry. They tended to have low education levels with the highest formal qualification being at the secondary level. These leaders typically ranked low on intellectual stimulation for employees and did not encourage creative thinking and innovation. This may probably be linked to the composition of their families where there was the absence of an avid teenage internet user whose example may have influenced perceptions. This trait has been deemed to be the most applicable transformational leadership trait to technology adoption research as it is the only one which appeals to cognitive change and an openness to creativity and new ideas. The researcher coined the term "intellectually stagnating" to signify the converse of intellectually stimulating. These respondents view internet technology adoption for sales and marketing as being a high-risk investment and activity and this meant that there was less proclivity towards adoption. These leaders were mainly found in firms which were categorized as computer adopters with their highest level of interface being the use of Global Distribution Systems.

Caretakers

The primary emphasis of this leader is to enforce the status quo and control activities to ensure adherence to guidelines. These leaders are transactional and also rank themselves as low risk-takers who will only do what is necessary to follow internal procedures regardless of external changes. This makes them just as intellectually stagnating as the previous group. They are only different from *resistors* to the extent that they are willing to do things differently if there is a complete industry change which warrants new procedures. Only after the development of new procedures will they alter any activity within the firm. They typically had low technology experience like *resistors* however their level of education tended to be a little higher as they had done some post-secondary studies such as certificates and diplomas. These were usually forms of certification from professional travel bodies. The moderate education level gave them exposure to technology benefits although they had limited experience using it. They however had no experience working outside of the travel industry and developed a limited view of business strategy. These leaders also ranked low on intellectual stimulation for employees and did not encourage creative thinking and innovation. Their family composition also tended to be void of an avid teenage internet user whose example may have influenced perceptions.

Fig. 5-3: Caretakers

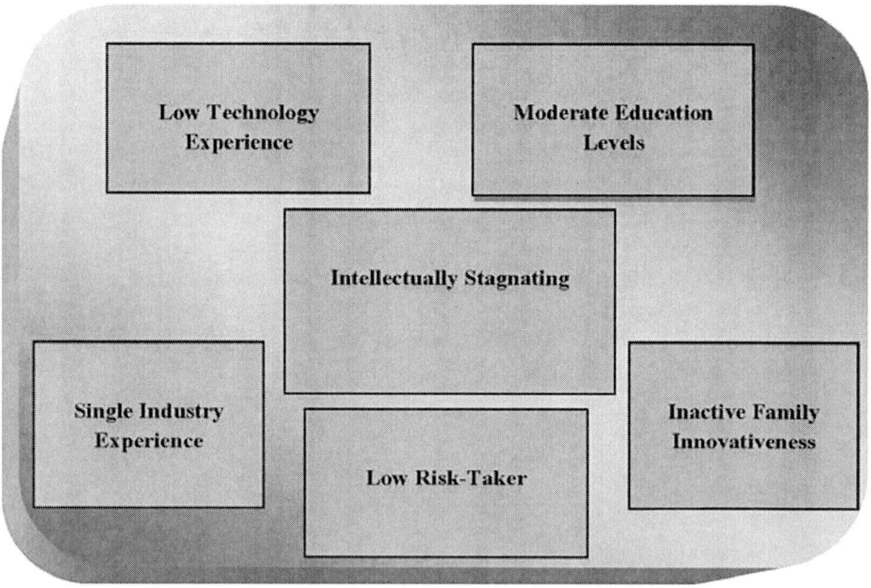

Despite changes in the global marketplace these leaders do not possess the strategic agility of a visionary and insist on upholding previous standards and procedures. These leaders were found in *internet adopter* firms and this was influenced by their greater exposure to greater levels of education than the *resistors*. Although they have adopted the internet, they were slow in doing so and only use it for basic emailing and web browsing.

Stabilizers

Fig. 5-4: Stabilizers

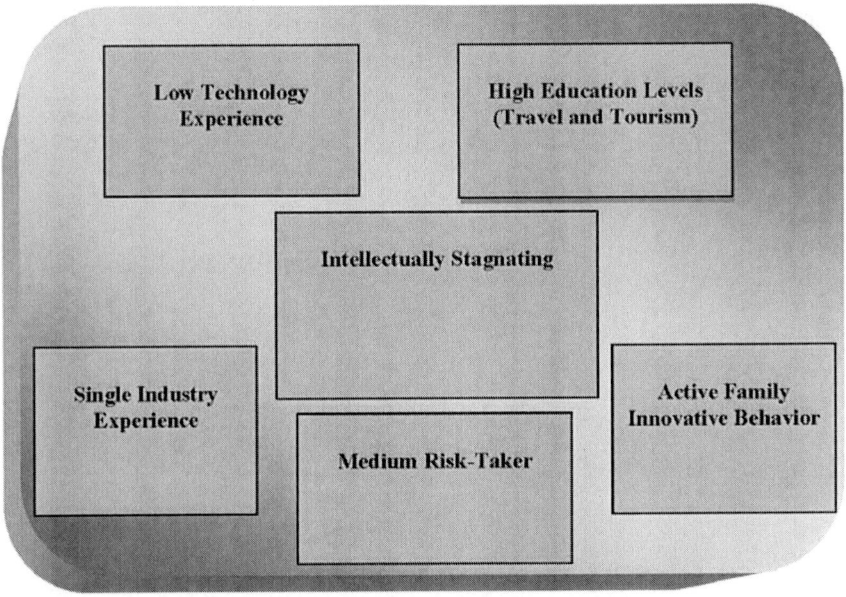

This type of leader is still a transactional leader but more closely resembles the *decision maker as actor* in the organizational decision-making literature. This is an individual who passively acts in accordance with what happens to the organization such as problems or opportunities. These leaders consider themselves to be medium risk-takers who will make decisions which may lead to short-term changes to address immediate opportunities and challenges. This may lead to doing something in the firm that is not procedural, however once the incident has passed it becomes business as usual once again.

They typically had low technology experience and fairly high education levels. This was usually a bachelor degree in travel/tourism which seemed to limit their scope. Their work experience was similar to *resistors* and *caretakers* in that they had only worked in the travel industry. Their family composition had teenagers using the internet which made them more familiar with its use but they were still no more intellectually stimulating than the previous categories as they usually tried to stabilize situations without employee consultation.

Evidence of this was seen where these leaders who were heads of *website adopter* companies introduced websites in response to competitors doing so without proper website design, orientation and training. The end result was that websites quickly became inactive and returns on investments were not realized. They had no intention of revitalizing their website efforts and did not intend to move any higher along the adoption hierarchy unless a crisis emerged.

Reactors

Fig. 5-5 Reactors

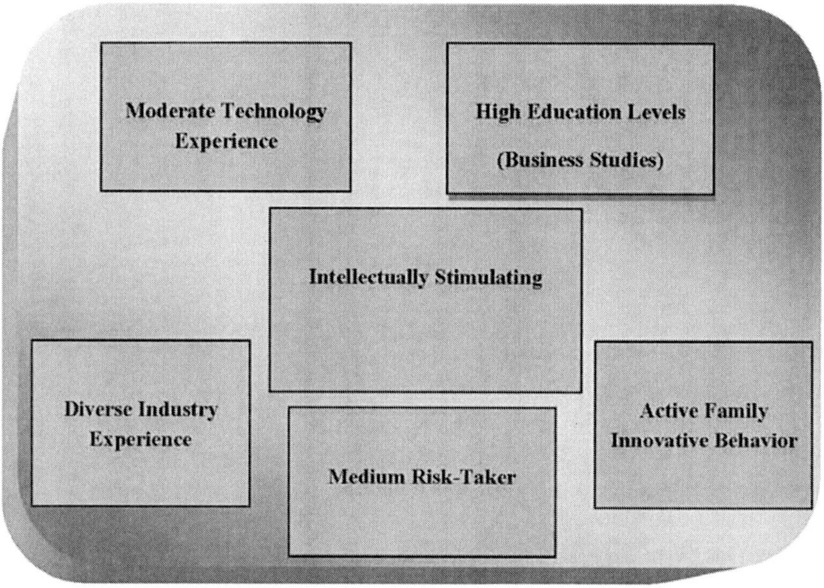

This category is representative of a leader who is much more open to ideas of change and is a medium risk-taker. Although not as resistant to doing things differently this leader is usually behind in the uptake of some new innovations. There is always the intention to improve business practices but these are sometimes late in coming to fruition. This is representative of the earlier discussed initiation phase and they many times get to implementation after much deliberation. Education levels are usually high in this category with a minimum of a bachelor degree with a broader focus than just travel and tourism. Technology experience is moderate (daily business emailing, and web browsing) and previous work experience involves working in another industry other than travel and tourism. They typically belong to families with avid teenage web users and provide an intellectually stimulating environment for employees but tend to spend a considerable time in the initiation phase of adoption.

This type of leader may move the firm to becoming e-commerce adopters, however sufficient research must take place especially since this will involve adopting a technology activity for the firm; one in which they have not been engaging personally. They differ from the *transformers* in that they are reactive rather than proactive. However when provided with sufficient stimulus they are open to change. Their reactive nature may be addressed through greater technology exposure and less risk aversion. A greater level of transformational leadership is needed to move the firm to relatively new business platforms such as social media. This will require a proactive visionary approach given that this activity is virtually non-existent in the context within which they operate.

Transformers

Considered to be an active change agent who has a long term vision for the firm, this visionary proactive leader is described as a *transformer* who is willing to change processes and approaches not only in response to external changes but also in anticipation of those changes. This is influenced by the fact that the individual is considered to be a high risk-taker with high technology experience and a high education level. The high technology experience and high risk-taking, which separate this leader from the *reactor*, are evidenced by the fact that this individual has engaged in personal online buying and is therefore familiar with online transactions. All other variables are similar to the *reactor* such as work experience, family composition and intellectual stimulation. The key difference is that in addition to being able to observe the benefits of use by their children, these leaders have their own experience with using cutting-edge technology. Although these leaders do not use social media in the businesses, they use it personally, which provides for an understanding of the vast reach of these tools.

Fig. 5-6: Transformers

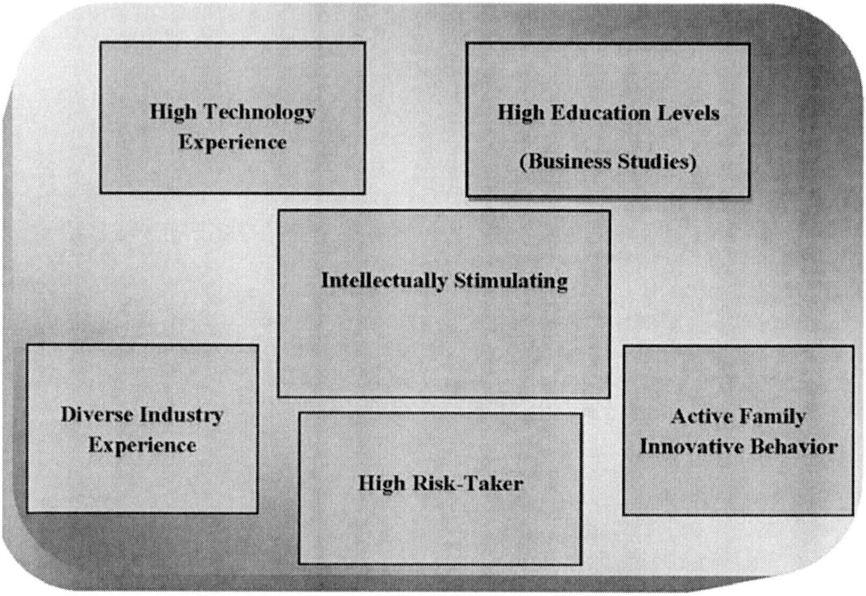

It must be noted that while this is the type of leader needed to move owner-managed small firms to the adoption of social media for sales and marketing, the leader in the sample who is most likely to move to this category is now only exploring the option and expressing an intention to adopt. This still makes the firm an industry leader in internet technology adoption for sales and marketing through the strategic vision and leadership of its top executive and owner. The composite model will now be illustrated and discussed below with an elaboration of the gaps being filled by this research.

The illustration in *Fig. 5-7* is a representation of this work's overall contribution to theory. It is an amalgamation of the previously explained leadership typologies and technology adoption hierarchy which emerged from this study. This composite diagram seeks to provide a coherent explanation for types of leaders which were found at various stages of adoption and highlight variables that separate each level. The model will assist in identifying potential adoption behaviour in small, owner-managed firms in information intensive industries and understanding what may influence change. While the study was carried out using travel firms, the implications are more far-reaching than that industry. Having discussed the construction of the model throughout this chapter, it is imperative that its strengths and limitations be discussed. The subsequent section will address the strengths of this study in relation to theoretical implications and the gaps in the body of literature which it will fill. Practical implications will also be highlighted. Inherent limitations of the model will also be presented and contrasted with how some level of delimitation is achieved. Ultimately the chapter will conclude with a reiteration and emphasis on the work's originality and theoretical contribution.

Strengths of the Model

The model presents a coherent hierarchical adoption approach which identifies types of leaders found at each level of adoption. Most importantly this model emerged out of a process of looking holistically at numerous possible factors that have been presented in previous research as drivers of technology adoption. This was facilitated through the lens of *Critical Social Science*, which called for a more comprehensive assessment of a multiplicity of possibilities.

As demonstrated in *figures 2.1* and *2.2*, factors such as culture, the digital divide, resources and firm strategy were also examined and secondary and primary data were collected on these factors in relation to technology adoption. The primary data analysis from two phases of collection and analysis revealed that culture and the digital divide only played an indirect role in the general shaping of individual's views but could not explain why similar cultural backgrounds and IT access resulted in different behaviours among the firms.

The resource and firm strategy factors emerged as peripheral drivers as they were directly under the control of the decision maker in these firms and appeared more as an effect than a cause. In other words resource and strategy decisions were the result of managerial direction rather than seriously influencing the decision to adopt.

The process of distillation in this research journey revealed that the critical direct driver of technology adoption in these firms was its leadership. This was primarily so because of the small autonomous nature of the firms in the study. This brings to the fore another important strength of the project. While there has been considerable research as previously discussed about technology adoption in large and small firms, owner-managed small firms have largely been ignored in the technology adoption discourse.

This book provides a model which emphasizes that where leaders of small firms are the owners themselves, this presents some unique challenges for innovative behaviour. First of all, the small size of the firms usually produces a simple management structure and hierarchy which allows for less bureaucracy and faster decision-making. This may on the surface appear to be a positive element however it also means that hasty decision-making at the whim of a leader's preferences, or lethargic decision-making based on a leader's dislikes are likely to become characteristic of the firm's behaviour.

The second issue is that as the owners of small firms these top executives have an immediate stake in the outcome of business activities. They are more likely to personalize decision-making rather than yield to the more strategic logic that will lead to greater long-term benefits. The bottom line of these two issues is that these leaders may be initiators or barriers to technology adoption based on their attitude to technology.

Having identified that the leadership element is most critical to technology adoption in owner-managed small firms, this research moved to a more detailed approach in identifying what firms were at various levels of adoption and what type of leader was responsible for the firm's position along the adoption hierarchy. While there has been previous research on stages of adoption, much of this research has been focused on the sequential adoption of specific technological innovations (see for example Cooper and Zmud, 1990; Damanpour, 1991; Daniel et al. 2002; Aquila-Obra and Padilla-Melendez, 2006). This research however takes a hierarchical approach which illustrates different technology adoption processes and therefore provides a macro approach to this conceptualization.

The major theoretical development is that the work goes on to develop leadership typologies at each level of adoption, which has never been done before. It must be stated that there were a few odd cases in the interviews which did not match the criteria of a particular typology based on the variables which were used to create the types of leaders. These were however in the minority and sufficiently resonant themes could be identified across interview manuscripts to make these bold conclusions. This takes the research a step further than previous research which identified the importance of leadership in innovation diffusion and adoption (see for example Thong and Yap, 1995; Peterson et al. 2009). Much of this body of work focused on the initial adoption of innovation in general or on a single innovation.

The development of leadership typologies is particularly useful as the generic transactional and transformational typologies were limited in their application to a hierarchical adoption model. For example the constructs identified in transformational leadership research did not directly apply to innovative behaviour with the exception of intellectual stimulation, which has been used in the development of this model. Additionally there were important variables which were not captured in these generic typologies but applied to the innovative behaviour of leaders such as technology experience and innovative family behaviour.

In the case of owner-managed small firms where the leadership element is most critical, it is of paramount importance that an applicable set of classifications which consider more technologically related variables be advanced in order to understand why some firms are at higher levels than others. This was particularly interesting in the travel context, where firms were faced with significant changes globally which could affect their level of importance. Even with impending danger to the survival of businesses, some firms were still at relatively low levels of adoption along the hierarchy and it was interesting to understand why this was so in such an information intensive industry. Although the study focused on travel firms, its contribution may be generalized to the extent that firms in other information intensive industries which are small and managed by their owners may learn lessons about leadership drivers and barriers to technology adoption.

The model provides useful information for small owner-managed firms in the travel industry as well as those with similar characteristics in other information intensive industries. The leadership typologies presented are instructive to firms aiming to become more competitive. They point to key variables which explain individual attitudes and behaviours regarding technology adoption in firms and can enable industry practitioners to aim for the highest level of transformational leadership and become as proactive as the *transformers* in anticipating competitive challenges and opportunities. Owner-managers of small travel firms are now able to explore their leadership style and traits through the lens of this study and assess whether their approaches to operations and strategy are directly linked to education, technology experience, work experience, family composition, risk aversion or level of intellectual stimulation (openness to new ideas from employees) which they provide in their firms. The ability to change one or more of these variables may result in a change in decision-making related to technological innovations and more specifically online selling.

Limitations of the Model

While the model makes a significant contribution to the understanding of leadership influence in levels of technology adoption, it does not do so without its share of limitations. The first is that although there is transferable potential in this work, the study was only conducted on small owner-managed firms in the travel industry for one country. Additionally in designing the leadership typologies, there were two outliers in the data set for two of the typologies.

The first issue may be addressed through the acknowledgement that the transferable nature of this work extends only to similar companies, similar industries and similar developing countries. Companies which are small, owner-managed and operating in information intensive industries tend to typify the large number of firms in developing countries.

The second more theoretically important issue is overridden by the resonance of the key themes emerging for the respondents in each leadership grouping. While there were differences in the variables for one respondent in each of two groups, a pattern in the overall responses was still clearly identified and the robustness of the findings was not compromised.

CONCLUSION

The findings of this research are original in its contribution in a number of ways. Firstly it posits that a holistic approach must be taken to research on technology adoption in firms, before key determinants are identified. This argument challenges the traditional reductionist approaches to technology adoption studies which have been guided by the positivist paradigm. An open, more inclusive approach, which is facilitated by *Critical Theory* allows for attendant or subliminal issues to be addressed before a process of distillation allows for a clearer perspective on critical emergent factors. A second important contribution is to the conceptualization of staged technology adoption through a synthesis of leadership characteristics and typologies that drive the movement of firms along a hierarchy of adoption. Much of the previous research identified sequential, horizontal stages of adoption for single innovations.

While this is not the only hierarchical representation of technology adoption, there is a certain degree of novelty in the coalescing of variables that drive movement to the next stage of adoption into clear typologies. Fundamentally this resulted in a valuable contribution to transformational leadership theory. The results revealed that while generic classifications of leadership have demonstrable value in explaining leaders' abilities as change agents, they were not sufficient to explain how leadership influenced staged technology adoption along an adoption hierarchy for small, owner-managed firms. The model in *Fig. 5-7* responds to this gap by further disaggregating these two generic typologies and creating smaller leadership typologies within the transactional/transformational paradigm that have greater application in the technology adoption discourse.

The findings of this research facilitated a re-conceptualization of the leadership element in the technology adoption discourse through an identification of leadership characteristics at each level of technology adoption. The key adoption being investigated was online selling practices; this was the highest level of adoption by any of the respondents in the study however only two of the firms were at this level. One firm was at the initiation phase of this adoption while the other was at the implementation phase. Social media adoption was also included as an aspirational level of adoption as none of the firms in the sample were using this tool for sales or marketing. The need therefore emerged to have a theory which guided the discussion of these relationships. Fundamentally technology adoption behaviour in these organizations related to how decisions were made about what technologies to adopt and how to implement and sustain their use. The theory of Organizational Decision-Making was therefore explored to gain insights into how decisions which affected the strategic future of organizations were conceptualized in previous research. This theory highlighted that the key to unravelling the complexity of decision making in organizations was to disaggregate the decision-making process in such a way that the individual decision-makers become the focus.

Through three major categorizations theorists identified the decision-maker as actor, creator and carrier. The actor refers to a decision-maker who is reactive and responds to situations while the creator is an innovative-minded individual who is proactive. The carrier is simply a decision-maker who brings previous experiences to the decision-making process. This latter category could also apply to actors and creators. Based on these conceptualizations *Fig. 1-2* was developed as a framework which demonstrated how decision-making took place in firms and how this ultimately affected technology adoption. While this was an output framework from the secondary data analysis it served as an input framework for the primary data collection.

The framework essentially posits that ownership as well as leadership, influenced by previous experiences, affects firm structure such as size, control and division. Also they influence creativity and risk-taking as elements of entrepreneurship and leadership issues such as motivation, stimulation, charisma and strategic change. The trickle-down effect is that strategy formulation, implementation and resource allocation are affected by these elements. Ultimately firm behaviour in terms of input, processes and output are influenced and this affects innovation adoption (Victorino et al. 2006).

The notion expressed earlier that an understanding of individual decision-making could demystify organizational decision-making was particularly resonant in this study where the entire sample involved owner-managed small firms. In the case of these firms decision-making did not involve complicated bureaucratic processes but rather simple personally driven choices that reflected the preferences of owners as demonstrated in their management focus. This differs from firms which do not have owner-managers, as the key variables such as technology experience, family composition and risk aversion would have had less significance in a firm which is not as dependent on the direction of one decision-maker. This highlighted the need to pay particular attention to ownership and leadership issues in this research.

The ownership and leadership constructs were coalesced in light of the fact that owner-managers were the subject of the study. Much of their leadership decisions were influenced by the fact that they had a personal stake in the business. In fact many of them were operating their businesses as an extension of themselves. The leadership element then became dominant in the investigation as it was seen to have a significant effect on adoption behaviour. This indicates that technology adoption depends on their overall ICT understanding.

The work has revealed that the perceptions of these owner-managers did not generally favour a greater use of internet technology in sales and marketing efforts. They typically felt that such technological investments were high-risk and yielded low returns on investments. Of particular interest was the fact that they were able to identify some potential benefits of using the internet such as speed, convenience and efficiency but this did not outweigh the perceived psychological risk of becoming too dependent on the internet. This they felt would mean that they would lose the strength of their personal interaction with clients. A few exceptional respondents had differing perceptions which led to a greater openness and intention to increase adoption. Their views suggested that the use of multiple platforms was the best approach. In this way personal contact would be maintained for those clients who required this, while new platforms could be created for more technologically savvy customers. These views were influenced by differences in education, technology experience, work experience, family background and risk aversion.

With owner-managers playing such a critical role in the decision-making processes of their firms, the need emerged to understand what shaped their perceptions and ultimately their leadership approach. A second phase of primary investigation focused on two dominant leadership types of transactional and transformational leadership through qualitative enquiry. Of the four key constructs from the most substantive work in the area of transformational leadership (Bass and Avolio, 2003), intellectual stimulation emerged as the most influential in the technology adoption discourse. This construct was measured through responses about a leader's openness to receiving new ideas from employees. All, with the exception of two respondents articulated that they did not encourage an environment in which employees could share ideas. They were more interested in employees simply focusing on daily tasks.

Individualized consideration, idealized influence and inspiration are all related to creating an atmosphere of admiration, respect and kinship to the leader. Although there was no difference between respondents in these areas, the intellectual stimulation construct showed that the more innovative-minded owner-managers were different and ranked higher in this area. An intellectually stimulating leader helps followers to become more innovative and creative by being open to their ideas. An exploration of the other three constructs was important to be able to clearly identify if these leaders would fit into any of these two leadership categories based on all four constructs. While the previous three demonstrate the likelihood of followers aligning themselves to the goals of the leader, the only one which addresses a cognitive and behavioral change in the follower is intellectual stimulation.

This meant that with only one construct having resonance, the general categories of transformational and transactional leadership were not sufficient to identify leaders which influenced various levels of technology adoption. While it is accepted in this work that it takes a transformational approach to move the firm higher up on the technology adoption hierarchy, a further disaggregation was needed to clearly demonstrate leadership typologies which emerged from the findings at each level of adoption. These typologies were a composite of the intellectual stimulation component (openness to employees' ideas) of transformational leadership as well as variables such as the level and type of formal education, technology experience, previous work experience, family composition and risk aversion.

An exploration of internal factors such as strategy and resources was necessitated by previous works which stated that a firm's resource capabilities (Barrney, 2001; Kraajenbrink e al. 2010) and its strategy decisions (David, 2007; Stonehouse and Snowdon, 2007) affect its behaviour. The factors were explored through secondary and primary data and the analysis reveals that throughout the process of distillation to identify the key driver of technology adoption for firms with these characteristics, strategy and resources were not central factors. They are not dismissed however as they are peripheral contributors in the sense that they are used by leaders to affect technology adoption processes but they do not provide the genesis for these processes. In many ways they emerge as a result of leadership behaviours and capability rather than drivers in and of themselves. They may turn out to be more influential in contexts which are different but for the purposes of small, owner-managed firms in information intensive industries they are more pawns in technology decisions rather than key determinants.

The exploration of these internal factors through primary data collection with all owner-managed travel firms in Jamaica revealed that there was no clear strategy which influenced technology adoption behaviour. It could be inferred that many firms essentially engaged in a hi-touch strategy (intense personal interaction) and feared that they could suffer from a perceived psychological risk of dependency on technology as posited by Eastlick and Lotz, (1999). They held the view that much of their clientele, as influenced by culture, was more inclined to place a high value on personal interaction. Additionally owner-managers felt that a hi-tech and hi-touch strategy were incompatible.

From a resource perspective owner-managers indicated that they face human and financial resource constraints. Many owner-managers implied however that they did not feel that there were other technologies which they needed but could not have because of limited resources. In simpler terms even if they had the resources they may not be inclined to adopt more technologies. The implication is that the decision of how to allocate already limited resources, are the sole purview of owner-managers. The tendency to prioritize spending in areas other than technology adoption was a reflection of the individual leader's preferences and comfort zone. Most respondents saw that there were potential benefits of greater technology adoption for their firms, however their own technology understanding and capability influenced adoption decisions. In the final analysis firm strategy and resources did not sufficiently explain differences in adoption behaviour in the firms. These factors were especially insignificant in explaining why some firms were at different levels of the technology adoption hierarchy as strategies and resource constraints were not heterogeneous enough to answer the question of why some firms were more technologically advanced than others. These internal factors emerged as a product of differences in the leadership of these small, owner-managed travel firms rather than as significant input factors into the decision to adopt newer technologies.

Contextual issues such as culture and the digital divide may explain overall industry behaviour but they are inadequate in explaining differences in behaviour among firms within the same society. The constructs assessed were values, norms, traditions and social interactions. The firms investigated are faced with similar external environments in which to operate and yet they make different choices operationally and strategically. The examination of culture as a factor revealed that firms were operating in a relationship oriented society where friendships and familiarity influenced business relationships. In addition to a relationship orientation, the national culture had an impact in that it promoted traditionalism with strict adherence to norms within the society. This had implications for firm behaviour as well as consumer behaviour and made it more difficult to make strategic decisions which required a deviation from these norms. The most important point however is that these cultural factors at the national level were experienced by all members of the sample which indicates that they may not be used to explain individual differences in firm adoption behaviour. A more determining element of culture stemmed from the sub-cultural influence of the family.

The composition of the family played a role in firm technology adoption through its influence on leaders of these firms. While the family background of employees may also have shaped their views it was deemed to be more important to assess its influence on leadership in the context of small owner-managed firms where leaders make unilateral decisions about innovation. It was found that family composition played a role especially in separating high adopters from those at the bottom of the adoption hierarchy.

Owner-managers who had family members living in the same household who were engaging in innovative behaviour were more open to innovative ideas and were more likely to introduce innovation in the firms. Although the influence of the family as a subculture explained larger disparities it did not offer an explanation for subtle differences between leadership categories. For example there were moderate adopters and high adopters who had family composition in common. This means that this variable must be used in combination with other variables to explain variances. The firms in the sample had equal access to technological resources and faced similar challenges in terms of the content and learning divides. This made the digital divide an even less significant context factor in explaining firm adoption behaviour.

A global divide is more evident than one which is internal to the country in the business context. In light of this fact these travel firms may reintroduce themselves as being relevant through bridging the technological divide (Minghetti and Buhalis, 2010) and communication gap (Maurer and Lutz, 2011) which exists between the destination and its more developed tourist generating countries such as the United States and the United Kingdom. This may be done through the use of covertly induced and autonomous agents such as the internet (Grovers et al. 2007).

An even amount of exposure to similar context factors means that their explanatory potential has been diminished in understanding technology adoption differences in the firms. Due to the fairly even influence of these factors on the firms within the society, they are instructive only to the extent that they can provide an enlightened approach to industry restructuring. This particular objective was primarily to determine whether these factors influence the individual adoption decisions of small owner-managed travel firms. Overwhelmingly these factors emerged as peripheral and culture only provided a sub-element of the key determinant which emerged as leadership.

Firms are essentially extensions of their individual leaders who are the owners of these firms and make all central and peripheral decisions about inputs, processes and outputs. These decisions typically reflect their individual preferences and experiences. This model as demonstrated in *Fig. 5-7* contributes to two main theoretical areas. The first is in the debate on stages of technology adoption. A key observation is that the body of research on stages of adoption usually takes a sequential continuum approach to understanding singular innovations. This is a useful approach, however much of it does not take into account multiple levels of adoption (see for example Daniel, 2002; Aquila-Obra and Padilla-Melendez 2006). For the purposes of this research there was a need to conceptualize some technology adoption stages which represented a more holistic demonstration of technology adoption within firms. As opposed to previous approaches this conceptualization does not only consider single innovations and accompanying sequential processes but uses a hierarchical approach to highlight different levels of adoption with each level representing a more cutting-edge type of adoption. This was based on the findings of the research for the firms in the sample. While this is not the main focus of this study, it was necessary to develop a hierarchy that corresponds to the new leadership typologies being advanced. Each stage of adoption refers to back office use and ultimately sales and marketing use at that level. These are stated in this research as computer adopters, internet adopters, website adopters, e-commerce adopters and social media adopters.

Along the hierarchy each stage is treated as a precondition for subsequent stages as emphasized in the innovation interdependence literature (Moital et al., 2009). Computer adopters are those firms which only use the computer technology for back office functions as well as GDS use. Internet adopters simply engage in emailing and web browsing. Website adopters are firms which use the online platforms for general and marketing information while e-commerce adopters transact payments using these platforms. Social media adopters refer to those firms using social media for sales and marketing. It must be noted that each level of adoption is conceptualized as being in an initiation phase or implementation phase.

The second and more profound contribution is the creation of new leadership typologies that apply specifically to these previously identified hierarchical levels of technology adoption in owner-managed small firms. The relationship between each leadership typology and level of adoption may be seen in *Fig. 5-7.* While there has been some research which has recognized the influence of leadership traits on small business adoption (see for example Thong and Yap, 1995; Peterson et al. 2009), this research failed to identify leadership characteristics as a driver of each stage of adoption. Rather the body of research has recognized leadership as contributory in overall IT adoption in small start-up firms.

The typologies which have emerged are referred to as resistors, caretakers, stabilizers, reactors and transformers. These typologies and the model in general are original in that this is the first work that has disaggregated generally accepted leadership categorizations to create a model that reflects leadership influence at each stage of a hierarchical technology adoption structure. While there were six key variables which influenced leadership, there were noticeable differences for which variables were most influential at each stage of adoption. The key leadership variable in moving the firm from computer adoption to internet adoption was education. Their education level was typically higher with leaders having done some post-secondary studies. For website adopters, the leadership variables which were most influential were education, risk-taking and family background. In comparison to the previous category they had university qualifications, active technology use at home and were medium risk-takers.

In order to move a firm from website adoption (general information) to ecommerce adoption the dominant variables were technology experience, industry experience and type of education. The leader typically had moderate technology experience involving daily emailing and browsing, had diverse industry experience and had university education which was not limited to travel and tourism studies. The key variables which needed improvement for those at the ecommerce level and would likely move the firms to social media adoption were technology experience and risk-taking. Ecommerce adopter firms had leaders who were medium risk-takers and had moderate technology experience. High technology experience may reduce the perception of risk or at the very least increase the likelihood of taking such a risk having experienced the benefits of using online commerce platforms. This may be inferred from the most innovative-minded respondent, who would still be at the ecommerce level, but who has intentions to explore social media options. This respondent already has high technology experience.

Most importantly this model emerged out of a process of looking holistically at various factors that have been presented in previous research as drivers of technology adoption. This was facilitated through the lens of *Critical Social Science*, which called for a more comprehensive assessment of a multiplicity of possibilities within the context of the study. The process of distillation in this research journey revealed that the critical direct driver of technology adoption in these firms was its leadership. This was primarily so because of the small autonomous nature of the firms. This brings to the fore another important strength of the project. While there has been considerable research as previously discussed about technology adoption in large and small firms, owner-managed small firms have largely been ignored in the technology adoption discourse.

In terms of theory the work seeks to inform future research on staged hierarchical adoption and leadership typologies. Technology adoption theories as well as leadership theories will now be influenced by this research. The more practical implications surround the travel and tourism industry significance in the context of struggling travel firms. There are some key areas of emphasis in this work that will inform the future of the technology adoption discourse. These key areas are advanced below.

- From a methodological standpoint the work argues against reductionist approaches to technology adoption studies and advocates for approaches such as those facilitated by a *Critical Social Science Philosophy*. This ultimately allows for attendant or subliminal issues to be addressed before a process of distillation allows for a clearer perspective on critical emergent factors.

- This work emphasizes that there are unique challenges for innovative behaviour in small firms where leadership and ownership coincide (*ie*. owner-managed firms).

- Size of the firm and the simplicity of the management and decision-making structure emphasize the leadership role.

- Owner-managers are personally attached to their businesses and rarely delegate key decisions such as the adoption of a new technology. Fundamentally this means that leaders in these firms will either stimulate adoption behaviour or restrict adoption behaviour. This work provides a more detailed analysis than previous works on specific leadership typologies that drive different levels of adoption in firms with similar external characteristics.

- The major theoretical development is that the work develops leadership typologies at each level of adoption, which has never been done before. The development of leadership typologies is particularly useful as the generic transactional and transformational typologies were limited in their application to a hierarchical adoption model. The implication is that in order for ICT adoption and competitiveness to be improved in small owner-managed travel firms, one or more of the identified leadership variables must be adjusted.

This also resulted in a valuable contribution to transformational leadership theory. The results revealed that while generic classifications of leadership have demonstrable value in explaining leaders' abilities as change agents, they were not sufficient to explain how leadership influenced staged technology adoption along an adoption hierarchy for small, owner-managed firms. In the final analysis this book has developed an applicable set of classifications which consider more technologically related variables to be advanced from a leadership perspective in order to understand why some firms are at higher levels than others. This model is transferable only to the extent that firms are small, owner-managed and operate in a common context as it relates to the digital divide and culture. This means that comparative research across cultures or societies may require different conceptualizations.

The findings of this research provide some useful insights for practitioners as well. This is particularly important in a context where travel agencies were facing significant challenges with drastic commission reductions that saw the number of firms decline from 105 to 43 over the period of 1999-2009. During that time there were very few changes to operational procedures in these firms over this period despite the changes which were taking place in the global market place. The most static area of operations related to technology use. Subsequent to the adoption of Global Distribution Systems (primarily Sabre and Amadeus) in the 1990s, there was very little new adoption of technologies in particular those which have a direct impact on sales and marketing.

There is a need for these firms, as well as firms facing similar conditions to assess the kind of leadership being provided. Based on the variables highlighted such as technology experience, education, type of industry experience, family composition, risk aversion and intellectual stimulation, owner-managed small firms may improve on variables which are controllable. This may include improving experience with technologies, education levels or the provision of an intellectually stimulating work environment. While this research agrees that the same level of technology adoption may not be suited for all firms it advocates that leaders should be less resistant to exploring ideas of a radical transformation of their business processes, which may include greater adoption. Ultimately such openness will yield more competitive strategies.

There are many areas for future research in this field. It is hoped that this research will inspire research that is more holistic in assessing the determinants of technology adoption. One key area which may be explored further is an investigation into the role of transformational leadership in technology adoption for large firms. Additionally this may be assessed in a developed country context to identify whether there are key differences. A comparative study across cultures may be insightful in uncovering the influence of culture and digital divide issues. Particularly within the tourism context there is room for greater research on leadership issues of firms in destination countries and firms from which their business tourists originate in more developed countries. Future research could benefit from quantitative analysis of the variables and typologies in this study in multiple environments.

BIBLIOGRAPHY

Adler, Ronald B., and Jeanne Elmhorst. 1999. Communicating at Work. 6th ed. Boston: McGraw-Hill.

Afuah, Allan, and Christopher Tucci. 2001. Internet Business Models and Strategies: Text and Cases. Boston: McGraw-Hill.

AIM. 2002. Austrian Internet Monitor 2002. Vienna: AIM.

—. 2003. Austrian Internet Monitor 2003. Vienna: AIM.

Akkermans-Allen, Hans. 2001. "Intelligent E-Business: From Technology to Value." IEEE Intelligent Systems 16:8-11.

Alba, Joseph, John Lynch, Barton Weitz, and Chris Janiszewski. 1997. "Interactive Home Shopping: Consumer, Retailer, and Manufacturer Incentives to Participate in Electronic Marketplaces." Journal of Marketing 61(3):38-53.

Alexander, Elmore R., Larry E. Penley, and I. Edward Jernigan. 1991. "The Effect of Individual Differences on Managerial Media Choice." Management Communication Quarterly 5(2):155-173.

Alford, Philip. 2004. "Critical Theory: An Alternative Solution to IT Planning Implementation Problems in Tourism." In ENTER 11: Information and Communication Technologies in Tourism Conference 2004: Proceedings of the International Conference in Cairo, Egypt, edited by Andrew J. Frew, 271-282. Vienna, Austria: Springer.

Allcock, Sue, Plenty Annette, Sheila Webber, and Robin Yeates. 1999. "Business Information and the Internet: Use of the Internet as an Information Resource for SMEs: Final Report: Section 18: Stages of Internet Engagement: Proposed Model and Recommendations." Accessed February 20, 2009. http://dis.shef.ac.uk/business/BII-18.pdf

Allen, Thomas J. 1977. Managing the Flow of Technology. Massachusetts: First MIT Press.

Al-Qirim, Nabeel. 2007. "A Research Trilogy into E-Commerce Adoption in Small Businesses in New Zealand." Electronic Markets 17(4):263-285.

Amit, Raphael, and Christoph Zott. 2001. "Value Creation in E-Business." Strategic Management Journal 22:493-520.

Andereck, Kathleen L., and Linda L. Caldwell. 1993. "The Influence of Tourists' Characteristics on Ratings of Information Sources for an Attraction." In Communication and Channel Systems in Tourism

Marketing, edited by Sheauhsing Hsieh, and Daniel R. Fesenmaier, 171-189. Gloucestershire: Hawthorn Press.

Anderson, Ronald, and David Reeb. 2003. "Founding Family-Ownership and Firm Performance from the S&P 500." Journal of Finance 58(3):1301-1327.

Andrews, Kenneth R. 1971. The Concept of Corporate Strategy. Homewood, IL.: Irwin.

Applegate, Lynda. 2001. "Emerging E-Business Models: Lessons from the Field." HBS 9(801): 172.

Aquila-Obra, Ana, and Antonio Padilla-Melendez. 2006. "Organizational Factors Affecting Internet Technology Adoption." Internet Research 16(1):94-110.

Assael, Henry. 1995. Consumer Behavior and Marketing Action. 3rd ed. Ohio: South-Western College Publishers.

Bagozzi, Richard, Fred Davis, and Paul Warshaw. 1992. "Development and Test of a Theory of Technological Learning and Usage." Human Relations 45(7):659-686.

Baker, Michael, and Jacob Reinders. 1997. The Future for Direct Retailing of Travel and Tourism Product the Influence of Information Technology. Edinburgh: Springer Verlag.

Baker, Elizabeth, Mary Barret, and Lesley Roberts. 2002. Working Communication. Queensland: John Wiley & Sons Australia.

Bakos, Yannis. 1998. "The Emerging Role of Electronic Marketplaces on the Internet." Communications of the ACM 41:66-42.

Baloglu, Seyhmus. 1999. "A Path Analytic Model of Visitation Intention Involving Information Sources, Socio-Psychological Motivations and Destination Image." Journal of Travel and Tourism Marketing 8(3):81-90.

Banff, Canada, Simon Hudson, and Norma Lang. 2002. "A Destination Case Study of Marketing Tourism Online." Journal of Vacation Marketing 8:155.

Barnes, James G. 1994. "Close to the Customer: But is it Really a Relationship?" Journal of Marketing Management 10:561-570.

Barnett, Martin, and Craig Standing. 2001. "Repositioning Travel Agencies on the Internet." Journal of Vacation Marketing 7:143.

Barney, Jay. 1991. "Firm Resources and Sustained Competitive Advantage." Journal of Management 17(1):99-120.

—. 2001. "Resource-based Theories of Competitive Advantage: A Ten Year Retrospective on the Resource-based View." Journal of Management 27:643.

Bass, Bernard M., Bruce J. Avolio, Dong I. Jung, and Yair Berson. 2003. "Predicting Unit Performance by Assessing Transformational and Transactional Leadership." Journal of Applied Psychology 88:207-218.

Bass, Bernard M. 1999. "Two Decades of Research and Development in Transformational Leadership." European Journal of Work and Organisational Psychology 8(1):9-32.

—. 1985. Leadership and Performance beyond Expectations. New York: Free Press.

Bear Stearns. 2000. Point, Click, Trip: An Introduction to the On-line Travel Industry. Bear Stearns & Co.

Beatty, Sharon E., and Scott M. Smith. 1987. "External Search Effort: An Investigation across Several Product Categories." Journal of Consumer Research 14:83-95.

Beck, Andrew, Peter Bennett, and Peter Wall. 2004. Communication Studies: The Essential Resource. London: Routledge.

Beck, Roman, Rolf Wigand, and Wolfgang Konig. 2005. "The Diffusion and Efficient Use of Electronic Commerce among Small and Medium-sized Enterprises: An International Three Industry Survey." Electronic Markets 15(1):38-52.

Beckinsale, Martin, Monder Ram, and Nicholas Theodorakopoulos. 2011. "ICT Adoption and Ebusiness Development: Understanding ICT Adoption amongst Ethnic Minority Businesses." International Small Business Journal 29:193-219.

Beekhuyzen, Jenine, Liisa Hellens, and Mark Siedle. 2005. "Cultural Barriers in the Adoption of Emerging Technologies." Accessed March 2, 2010. http://www.ucd.smartinternet.com.au/Documents/Cultural_Barriers.pdf

Bei, Lien-Ti, Etta Y. I. Chen, and Richard Widdows. 2004. "Consumers' Online Information Search Behavior and the Phenomenon of Search vs. Experience Products." Journal of Family and Economic Issues 25(4):449-467.

Beirne, Elizabeth, and Philip Curry. 1999. "The Impact of the Internet on the Information Search Process and Tourism Decision-Making." In Information and Communication Technologies in Tourism 1999: Proceedings of the International Conference in Innsbruck, edited by Dimitrios Buhalis, and Walter Schertler, 88-97. Wien: Springer.

Belch, George E., and Michael A. Belch. 2001. Advertising and Promotion: An Integrated Marketing Communications Perspective. New York: McGraw Hill.

Beliveau, Annie, and John B. Garwood. 2001. "The Impact of New Technologies on Information Research and Purchasing Behaviour of

Quebec Air Travellers." In Information and Communication Technologies in Tourism 2001: Proceedings of the International Conference in Montreal, Canada, edited by Pauline Sheldon, Karl Wober, and Daniel Fesenmaier, 310-316. Wien: Springer.

Belk, Russell W. 1974. "An Exploratory Assessment of Situational Effects in Buyer Behavior." Journal of Marketing Research 11:156-163.

—. 1975. "Situational Variables and Consumer Behavior." Journal of Consumer Research 2:157-163.

Benjamin, Robert, and Rolf Wigand. 1995. "Electronic Markets and Virtual Value Chains on the Information Superhighway." Sloan Management Review 36:62-72.

Bennett, Marion, and Chi-Wen Lai. 2005. "The Impact of the Internet on Travel Agencies in Taiwan." Journal of Tourism and Hospitality Research 6(1):1-17.

Berle, Adolf, and Gardiner Means. 1932. The Modern Corporation and Private Property. New York: Harcourt, Brace and World.

Berlo, David K. 1960. The Process of Communication. San Francisco: Holt, Reinhart and Winston.

Bhaskar, Roy. 1989. Reclaiming Reality. London: Verso.

Bhimani, Alnoor. 1999. "Mapping Methodological Frontiers in Cross-National Management Control Research." Accounting, Organizations and Society 24:413–40.

Bieger, Thomas, and Christian Laesser. 2001. "Segmenting Travel on the Sourcing of Information." In Consumer Psychology of Tourism, Hospitality and Leisure, edited by Josef A. Mazanec, 153-168. Wallingford: CABI.

Bigne, Enrique J., Joaquín Aldas, and Luisa Anderu. 2008. "B2B Services: IT Adoption in Travel Agency Supply Chains." Journal of Services Marketing 22(6):454- 464.

Bloch, Michael, and Arie Segev. 1996. "The Impact of Electronic Commerce on the Travel Industry: An Analysis Methodology and Case Study." Hawaii International Conference on System Sciences (HICSS) 4:48-58.

Bloch, Michael, and Yves Pigneur. 1996. On the Road of Electronic Commerce: A Business Value Framework, Gaining Competitive Advantage. Berkeley: The Fisher Center for Information Technology & Management.

Bloch, Peter, Daniel Sherrell, and Nancy Ridgway. 1986. "Consumer Search: An Extended Framework." Journal of Consumer Research 13:119-126.

Bowersox, Donald, and David Class. 1996. Logistical Management: The Integrated Supply Chain Process. New York: McGraw Hill Companies.

Brewer, John, and Albert Hunter. 1989. Multimethod Research: A Synthesis of Styles. Newbury Park: Sage Publications.

Brouthers, Keith D., Lance E. Brouthers, and Steve Werner. 2008. "Resource-based Advantages in the International Context." Journal of Management 34(2):189.

Brown, James, Chris Hendry, and Paul Harborne. 2007. "Developing Radical Technology for Sustainable Energy Markets: The Role of New Small Firms." International Small Business Journal 25:603- 625.

Brown, Lawrence. 1981. Innovation Diffusion: A New Perspective. London: Methuen.

Buhalis, Dimitrios. 1994. "Information and Telecommunications Technologies as a Strategic Tool for Small and Medium Tourism Enterprises in the Contemporary Business Environment." In Tourism the State of the Art: The Strathclyde Symposium, edited by A. V. Seaton, 254-275. England: Wiley and Sons.

—. 1996. Information and Telecommunication Technologies as a Strategic Tool for Tourism Enhancement at Destination Regions. Innsbruck: Springer Verlag.

—. 1997. Catalysts in Introducing Information Technology in Small and Medium Sized Hospitality Organizations. Edinburgh, Springer Verlag.

—. 1998. Strategic Use of Information Technologies in the Tourism Industry. Elsevier Science.

—. 2000. "Tourism and Information Technologies: Past, Present and Future." Tourism Recreation Research 25(1):41-58.

—. 2001. "Tourism Distribution Channels: Practices, Issues and Processes." In Tourism Distribution Channels: Practices, Issues and Transformations, edited by Dimitrios Buhalis, and Eric Laws, 7-32. New York: Continuum.

—. 2002. eTourism: Information Technologies for Strategic Tourism Management. Essex: Pearson Education.

—. 2004. "E-Airlines: Strategic and Tactical Use of ICT's in the Airline Industry." Information and Management 41(7):805-825.

Buhalis, Dimitrios, and Ourania Deimezi. 2004. "E-tourism Developments in Greece: Information Communication Technologies Adoption for Strategic Management of the Greek Tourism Industry." Tourism and Hospitality Research 5(2):1467-3584.

Buhalis, Dimitrios, and Eric Laws. 2001. Tourism Distribution Channels: Practices, Issues and Transformations. London: Continuum.

Buhalis, Dimitrios, and Maria Cristina Licata. 2002. "The Future eTourism Intermediaries." Journal of Tourism Management 23:207-220.

Burgess, Lois, and Joan Cooper. 1998. "The Status of Internet Commerce in the Manufacturing Industry in Australia: A Survey of Metal Fabrication Industries." Collector 65-73.

Burns, James M. 1978. Leadership. New York: Harper and Row.

Cai, Liping. A., Ruomei Feng, and Deborah Breiter. 2004. "Tourist Purchase Decision Involvement and Information Preferences." Journal of Vacation Marketing 10(2):138-148.

Carland, James W., Frank Hoy, William R. Boulton, and Jo Ann Carland. 1984. "Differentiating Entrepreneurs from Small Business Owners: A Conceptualization." Academy of Management Review 9:354- 359.

Capella, Louis M., and Alan J. Greco. 1987. "Information Sources of Elderly for Vacation Decisions." Annals of Tourism Research 14 (1): 148-151.

Castells, Manuel. 2000. End of Millennium. 2nd ed. Oxford: Blackwell.

Catchcart, Robert, and Gary Gumpert. 1983. "Mediated Interpersonal Communication." Quarterly Journal of Speech 69(2):267-277.

Caves, Richard E. 1980. "Industrial Organization, Corporate Strategy and Structure." Journal of Economic Literature 58:64-92.

Central Information Technology Office. 2007. "National e-Readiness Status." Accessed May 12, 2011. http://www.cito.gov.jm/content/national-e-readiness-status

Chan, Caroline, and Paula Swatman. 2004. "B2B E-commerce Stages of Growth: The Strategic Imperatives." In 37th International Conference on System Sciences, 1-10. Hawaii.

Chandler, Alfred. 1963. Strategy and Structure: Chapters in the History of Industrial Enterprise. Cambridge: MIT Press.

Chandler, Alfred D. 1977. The Visible Hand: The Managerial Revolution in American Business. Cambridge: Belknap/Harvard University Press.

Chen, Joseph S. A. 2000. "Comparison of Information Usage between Business and Leisure Travelers." Journal of Hospitality and Leisure Marketing 7(2):65-75.

Chen, Joseph. S., and Dogan Gursoy. 2000. "Cross-cultural Comparison of the Information Sources used by First-time and Repeat Travelers and its Marketing Implications." International Journal of Hospitality Management 19:191-203.

Chen, Jihong, and Robert McQueen. 2008. "Factors Affecting E-commerce Stages of Growth in Small Chinese firms in New Zealand:

An Analysis of Adoption Motivators and Inhibitors." Journal of Global Information Management 16(1):26-60.

Cheyne, Jo, Mary Downes, and Stephen Legg. 2006. "Travel Agent vs. Internet: What Influences Travel Consumer Choices?" Journal of Vacation Marketing 12(1):41-57.

Childers, Thomas. 1975. The Information Poor in America. Lanham, MD: Scarecrow.

Christopher, Martin. 1986. The Strategy of Distribution Management. Oxford: Heinemann Professional Publishing.

—. 1992. Logistics and Supply Chain Management: Strategies for Reducing Cost and Improving Service. London: Prentice Hall.

Choa, Myeong-Hyeon. 1998. "Ownership Structure, Investment and the Corporate Value: An Empirical Analysis." Journal of Financial Economics 47(1):103-121.

Choi, Soon-Yong, and Dale Stahl. 1997. The Economics of Electronic Commerce. Indiana: Macmillan Technical Publishing.

Chon, Kye-Sung. 1999. "Tourist Information-seeking Behaviour and its Implications." Hospitality Research Journal 14:485-490.

Chorn, Norman. 1994. "The New Logistics Organization." In The Gower handbook of Logistics and Distribution Management. 4th ed., edited by John Gattorna, 501-508. Aldershot, England: Gower Publishing.

Christensen, Edward W., and James Bailey. 1997. "A Source Accessibility Effect on Media Selection." Management Communication Quarterly 10(3):373-387.

Christopher, Martin. 1998. Logistics and Supply Chain Management: Strategies for Reducing Cost and Improving Service. London: Prentice Hall.

Churchill, Gilbert A. 1995. Marketing Research. Chicago: Dryden Press.

Clark, Peter, and Richard Blundel. 2007. "Critical Realism and the Evolution of Business Knowledge: A Methodological Reappraisal." Management and Organizational History 2(1):45-62.

Cockburn, Claire, and Thomas Wilson. 1996. "Business Use of the World Wide Web." International Journal of Information Management 16(2):83-102.

Cole, Rebel A., and Hamid Mehran. 1998. "The Effect of Changes in Ownership Structure on Performance: Evidence from the Thrift Industry." Journal of Financial Economics 50(3):291-317.

comScore. 2002. "Online Travel Launches First Quarter 2002 E-commerce Sales to Record $17 Billion Level." comScore Press release. Accessed February 13, 2010. www.comscore.com.

Conlisk, John. 1996. "Why Bounded Rationality?" Journal of Economic Literature 34(2):669-700.

Connolly, Daniel J., Michael D. Olsen, and Richard Moore. 1998. "The Internet as a Distribution Channel." Cornell Hotel & Restaurant Administration Quarterly 39(4):42-54.

Cook, Suzanne. 2001. Travel Planners: How They Use the Web Today. Montreal: Springer Verlag.

Cooper, Randolph B., and Robert Zmud. 1990. "Information Technology Implementation Research: A Technological Diffusion Approach." Management Science 36(2):123-39.

Corey, Robert J. 1996. "A Drama-based Model of Traveler Destination Choice." Journal of Travel and Tourism Marketing 5(4):1-22.

Covin, Jeffery G., and Dennis Slevin. 1988. "The Influence of Organization Structure on the Utility of an Entrepreneurial Top Management Style." Journal of Management Studies 25(3): 217–234.

Cox, Andrew, and Mike Townsend. 1998. Strategic Procurement in Construction. London: Thomas Telford.

Cox, Andrew. 1998. "Relational Competence and Strategic Procurement Management: Towards an Entrepreneurial and Contractual Theory of the Firm." European Journal of Purchasing and Supply Management 2:57-70.

Crabtree, Benjamin, and William Miller. 1999. "A Template Approach to Text Analysis: Developing and Using Codebooks." In Doing Qualitative Research, edited by Benjamin Crabtree, and William Miller, 163-177. Newbury Park, CA: Sage.

Craig, Anita P. 2003. "Culture and the Individual." Theory and Psychology 13(5):629–650.

Cresswell, John W. 2006. Research Design: Qualitative and Quantitative Approaches. California: Sage Publications.

Creswell, John W., Vicki L. P. Clark, Michelle L. Gutmann, and William E. Hanson. 2003. "Advanced Mixed Methods Designs." In Handbook of Mixed Methods in Social and Behavioral Research, edited by Abbas Tashakkori, and Charles Teddlie, 209-240. London: SAGE Publications.

Crotts, John, Abdul Aziz, and Andrew Raschid. 1998. "Antecedents of Supplier's Commitment to Wholesale Buyers in the International Travel Trade." Tourism Management 19(2):127-134.

Crotts, John C. 1999. "Consumer Decision Making and Pre-purchase Information Search." In Consumer Behavior in Travel and Tourism, edited by Abraham Pizam, and Yoel Mansfeld, 149-168. New York: Haworth Press.

Cyert, Richard M., and James March. 1963. A Behavioural Theory of the Firm. Englewood Cliffs, NJ: Prentice Hall.

Daft, Richard L., and Robert Lengel. 1984. "Information Richness: A New Approach to Managerial Behavior and Organizational Design." Research in Organizational Behavior 6:191-233.

Dale, Crispin. 2003. "The Competitive Networks of Tourism E-mediaries: New strategies, New Advantages." Journal of Vacation Marketing 9:109.

D'Ambra, John, and Concepcion Wilson. 2004. "Use of the World Wide Web for International Travel: Integrating the Construct of Uncertainty in Information Seeking and the Task- technology Fit (TTF) Model." Journal of the American Society for Information Science and Technology 55 (8):731-742.

D'Iribarne, Philippe. 2009. "National Cultures and Organisations in Search of a Theory: An Interpretative Approach." International Journal of Cross Cultural Management 9:309-321.

Dahles, Heidi, and Karin Bras. 1999. "Entrepreneurs in Romance: Tourism in Indonesia." Annals of Tourism Research 26:267-293.

Daily, Catherine M., and Marc Dollinger. 1992. "An Empirical Examination of Ownership Structure in Family and Professionally-managed Firms." Family Business Review 5(2):117-36.

Damanpour, Fariborz. 1991. "Organizational Innovation: A Meta-analysis of Effects of Determinants and Moderators." Academy of Management Journal 34(3):555-90.

Dance, Frank E. X. 1967. "A Helical Model of Communication." In Human Communication Theory, edited by Frank Dance. New York: Holt, Reinhart and Winston.

Daniel, Elizabeth, Hugh Wilson, and Andrew Myers. 2002. "Adoption of E-commerce by SMEs in the UK: Towards a Stage Model." International Small Business Journal 20:253-270.

Dann, Graham M. 1996. "Tourists' Images of a Destination - An Alternative Analysis." Journal of Travel and Tourism Marketing 5(1/2):41-55.

Danzinger, Claudia, and H. Nowak. 2002. Internetnutzung und Ecommerce Aktivitäten der österreichischen KMU. Vienna: Ec3.

David, Fred. 2007. Strategic Management: Concepts and Cases. New Jersey: Prentice Hall.

Davies, Omar. 1986. "An Analysis of the Management of the Jamaican Economy: 1972-1985." Social and Economic Studies 35(1):73-110.

Davis, Fred D. 1989. "Perceived Ease of Use, and User Acceptance of Information Technology." MIS Quarterly 13(3):319-339.

Decrop, Alain. 1999. "Triangulation in Qualitative Tourism Research." Tourism Management 20: 157-161.

Deeds, David, Paul Mang, and Michael Fransden. 2004. "The Influence of Firms' and Industries' Legitimacy on the Flow of Capital into High-technology Ventures." Strategic Organization 2(1): 9–34.

DeFleur, Melvin L. 1966. Theories of Mass Communication. New York: David McKay.

DeHayes, Daniel W., and Robert L. Taylor. 1972. "Making Logistics Work in a Firm." Business Horizons 24(1):1-26.

Dellaert, Benedict G. C. 1999. "The Tourist as a Value Creator on the Internet." In Information and Communication Technologies in Tourism 1999: Proceedings of the International Conference in Innsbruck, Austria, 1999, edited by Dimitrios Buhalis, and Walter Schertler, 66-76. Wien: Springer.

DeMartino, Richard, Robert Barbato, and Paul Jacques. 2006. "Exploring the Career/Achievement and Personal Life Orientation Differences between Entrepreneurs and Non Entrepreneurs: The Impact of Sex and Dependents." Journal of Small Business Management 44(3):350-369.

Demsetz, Harold, and Kenneth Lehn. 1985. "The Structure of Corporate Ownership." Journal of Political Economy 93 (6):1155-77.

Denzin, Norman K., and Yvonna Lincoln. 1994. "Entering the Field of Qualitative Research." In Handbook of Qualitative Research, edited by Norman Denzin, and Yvonna Lincoln, 1-17. Thousand Oaks.

Desinano, Paolo, and Carlo Vigo. 1994. Developing Information Technology Options in the Hotel Industry: The Value Chain Approach. Tourism - the State of Art. New York: John Wiley & Sons.

Dettmer, Harald. 1998. Tourismuswirtschaft. Köln: Wirtschaftsverlag Bachem.

Dietrich, Brenda L., Jane L. Snowdon, and JoAnn Washam. 1997. "The Promise of Information Technology in the Travel Industry." In Information and Communication Technologies in Tourism: Proceedings of the International Conference in Edinburgh, Scotland, 1997, edited by A. Min Tjoa, 129-139. Wien: Springer.

Dodd, Tim H. 1998. "Influences on Search Behaviour of Industrial Tourists." Journal of Hospitality and Leisure Marketing 5(2/3):77-94.

Dombey, Alyson. 1998. "Separating the Emotion from the Fact: The Effects of the New Intermediaries on Electronic Travel Distribution." In Information and Communication Technologies in Tourism 1998: Proceedings of the International Conference in Istanbul, Turkey, 1998, edited by Dimitrios Buhalis, A. Min Tjoa, and Jafar Jafari, 129-138. Wien: Springer.

Downes, Larry, and Chunka Mui. 1998. Unleashing the Killer App. Boston: Harvard Business School Press.

Drori, Gili, and Yong Suk Jang. 2003. "The Global Digital Divide: A Sociological Assessment of Trends and Causes." Social Science Computer Review 21:144-161.

Drucker, Peter. 1990. Post Capitalist Society. New York: Harper Business.

Duke, Charles R., and Margaret Persia. 1993. "Effects of Distribution Channel Level on Tour Purchasing Attributes and Information Sources." In Communication and Channel Systems in Tourism Marketing, edited by Daniel Fesenmaier, and Muzaffer Uysal, 37-55. Gloucestershire: The Hawthorn Press.

Dulle, Frankwell W., and M.K. Minishi-Majanja. 2011. "The Suitability of the Unified Theory of Acceptance and Use of Technology (UTAUT) Model in Open Access Adoption Studies." Information Development 27(1):32–45.

Durand, Rodolphe, and Vicente Vargas. 2003. "Ownership, Organization, and Private Firms' Efficient Use of Resources." Strategic Management Journal 24(7):667–75.

Dyerson, Romano, and G. Harindranath. 2007. "ICT Adoption and Use by SMEs in the UK: A Survey of South East." Paper Presented at the Portland International Center for Management of Engineering and Technology (PICMET), Portland, Oregon - USA.

Eames, Donald, and Gregory Norkus. 1988. "Developing Your Procurement Strategy." Cornell Hotel and Restaurant Administration Quarterly 29:30-33.

Earl, Michael. 2000. "Evolving the E-business." Business Strategy Review 11(2):33-38.

Eason, Ken. 1988. Information Technology and Organizational Change. London: Taylor & Francis.

Eastlick, Mary Ann, and Sherry Lotz. 1999. "Profiling Potential Adopters and Non-adopters of an Interactive Electronic Shopping Medium." International Journal of Retail and Distribution Management 27(6):209-223.

Eastman, Richard. 1996. "The Future of Travel Agents." Travel Weekly 29.

Eby, David W., Lisa J. Molnar, and Liping A. Cai. 1999. "Content Preferences for In-vehicle Tourist Information Systems: An Emerging Travel Information Source." Journal of Hospitality and Leisure Marketing 6(3):41-58.

EC. 1999. ISTAG Report - Orientations for Work Programme 2000 and Beyond. Brussels: European Commission.

—. 2003. The European Ebusiness Report 2002/2003. Brussels: European Commission.

Elenkov, Detelin, and Ivan Manev. 2005. "Top Management Leadership and Influence on Innovation: The Role of Socio-cultural Context." Journal of Management 31(3):381-402.

Ellard, Al, and Norma Nickerson. 1999. Navigating Cyberspace: Information Seeking Strategies and Travel Behavior of Montana Vacation Visitors. TTRA Annual Conference.

Elliott, Dawn R. 2000. "Underdevelopment in Jamaica: An Institutionalist Perspective." Journal of Economic Issues 34.

Engel, James F., Martin Warshaw, and Thomas Kinnear. 1991. Promotional Strategy. 7th ed. Boston: Irwin.

Engel, James F., Roger D. Blackwell, and Paul W. Miniard. 1995. Consumer Behaviour. 8th ed. Fort Worth, TX: Dryden.

Enright, Michael J., and James Newton. 2005. "Determinants of Tourism Destination Competitiveness in Asia Pacific: Comprehensiveness and Universality." Journal of Travel Research 43:339-350.

Etzel, Michael J., and Russell G. Wahlers. 1985. "The Use of Requested Promotional Materials by Pleasure Travelers." Journal of Travel Research 23:2-6.

Evans, Philip B., and Thomas S. Wurster. 1997. "Strategy and the New Economics of Information." Harvard Business Review 75(5):70-82.

Fall, Lisa T. 2000. Segmenting Pleasure Travelers on the Basis of Information Sources Usefulness and Personal Value Importance. Michigan State University.

Falk, Martin. 2008. "Effects of Foreign Ownership on Innovation Activities: Empirical Evidence for Twelve European Countries." National Institute Economic Review 204:85-97.

Feather, John. 1998. The Information Society: A Study of Continuity and Change. 2nd ed. London: Library Association Publishing.

Fesenmaier, Daniel R., and Christine Vogt. 1992. "Evaluating the Utility of Touristic Information Sources for Planning Midwest Vacation Travel." Journal of Travel and Tourism Marketing 1(2): 1-18.

Fesenmaier, Daniel R., and Stefan Klein. 2000. Information and Communication Technologies in Tourism. Barcelona, Springer Verlag.

Fesenmaier, Daniel R., and Jiann-Min Jeng. 2000. "Assessing the Structure in the Pleasure Trip Planning Process." Tourism Analysis 5(1):13-29.

Fischer, Ronald. 2009. "Where is Culture in Cross Cultural Research?: An Outline of a Multilevel Research Process for Measuring Culture as a

Shared Meaning System." International Journal of Cross Cultural Management 9:25-49

Fishbein, Martin. 1967. Readings in Attitude Theory and Measurement. New York: Wiley.

Fishbein, Martin, and Icek Azjen. 1975. Belief, Attitude, Intention and Behaviour: An Introduction to Theory and Research. Reading, MA: Addision-Wesley.

Fishbein, Martin, and Icek Ajzen. 1980. "Predicting and Understanding Consumer Behavior: Attitude-behavior Correspondence." In Understanding Attitudes and Predicting Social Behavior, edited by Martin Fishbein, and Icek Ajzen, 148-156. Englewood Cliffs: Prentice-Hall.

Flognfeldt, Thor, and Ragnar Nordgreen. 1999. Information Used for Decision of Making Temporary Stops along Tourist Routes in Norway. Halifax, Canada: Annual TTRA Conference.

Fodness, Dale, and Brian Murray. 1997. "Tourist Information Search." Annals of Tourism Research 24(3):503-523.

Fodness, Dale, and Brian Murray. 1998. "A Typology of Tourist Information Search Strategies." Journal of Travel Research 37(2):108-119.

Fodness, Dale, and Brian Murray. 1999. "A Model of Tourist Information Search Behaviour." Journal of Travel Research 37(3):220-230.

Fontayne, Cynthia. 1996. The Connected Marketer: Using the Internet, World Wide Web and On-line Services for Marketing Travel. Madrid: Tourism and New Information Technologies.

Forman, Andrew M., and Ven Sriram. 1991. "The Depersonalization of Retailing- Its Impact on the "Lonely" Customer." Journal of Retailing 67:226- 243.

Forrester. 1998. "European New Media Strategies." Forrester Research 16.

—. 1999. "On-line Retail Strategies." Forrester Research 15.

Frese, Michael, Marco van Gelderen, and Michael Ombach. 2000. 'How to Plan as a Small Scale Business Owner: Psychological Process Characteristics of Action Strategies and Success" Journal of Small Business Management 38:1–18.

Freudenthaler, Johannes, and Wilfried Grossmann. 2002. Data Warehouse Models for Web-data. EC3. Vienna: University of Vienna.

Frew, Andrew J. 2000. "Information and Communications Technology Research in the Travel and Tourism Domain: Perspective and Direction." Journal of Travel Research 39(2):136-145.

Frey, Susanne, and Schegg Roland. 2000. Bench Marketing Internet Use for Marketing of Swiss Hotels. Innsbruck: Springer Verlag.

Fuchs, Matthias, Wolfram Hopken, Andreas Foger, and Martin Kunz. 2009. "E-business Readiness, Intensity and Impact: An Austrian Destination Management Organization Study." Journal of Travel Research 10.

Fulk, Janet, Joseph Schmitz, and Charles W. Steinfeld. 1990. "A Social Influence Model of Technology Use." In Organizations and Communications Technology, edited by Janet Fulk, and Charles Steinfeld, 117-140. Newbury Park, Cal.: Sage Publications.

Gage, Nathaniel. 1989. "The Paradigm War and Their Aftermath: A "Historical" Sketch Research on Teaching." Education Researcher 18:4-10.

Gallo, Peter J., and Lisa Jones Christensen. 2011. "Firm Size Matters: An Empirical Investigation of Organizational Size and Ownership on Sustainability-related Behaviors." Business Society 50: 315-349.

Gandhi, Sunil. 2006. "E-commerce and Information Technology Act, 2000." Vidyasagar University Journal of Commerce 11:82-91.

Gartner Group. 2000. "Gartner Group's Online Travel Forecast." Accessed October 4, 2010. http://gartner12.gartnerWeb.com/public/static/about gg/pressrel/pr010600.ht

Gatautis, Rimantas, and Bronius Neverauskas. 2005. "E-commerce Adoption in Transition Economies: SMEs Perspectives in Lithuania." In 7th International Conference on Electronic Commerce, 109-113. Xi'an, China: ACM New York, NY, USA.

Gattorna, John. 1994. The Gower Handbook of Logistics and Distribution Management. 4th edition. Aldershot, England: Gower Publishing.

Gatty, Bob. 1998. "E-Commerce Gives Lodging Industry Edge." Hotel and Motel Management 213.

Gelade, Garry, Paul Dobson, and Patrick Gilbert. 2006. "National Differences in Organizational Commitment: Effect of Economy, Product of Personality, or Consequence of Culture?" Journal of Cross-Cultural Psychology 37(5):542-556.

George, Gerard, Johan Wiklund, and Shaker Zahra. 2005. "Ownership and the Internationalization of Small Firms." Journal of Management 31(2):210-233.

Ghobadian, Abby, and Nicholas O'Regan. 2006. "The Impact of Ownership on Small Firm Behaviour and Performance." International Small Business Journal 24(6):555-584.

Ghosh, Shikhar. 1998. "Making Business Sense of the Internet." Harvard Business Review 124 - 135.

Gibson, Cristina, and Dana McDaniel. 2010. "Moving Beyond Conventional Wisdom Advancements in Cross-cultural Theories of

Leadership, Conflict, and Teams." Perspectives on Psychological Science 5:450-462.

Gitleson, Richard J., and John Crompton. 1983. "The Planning Horizons and Sources of Information Used by Leisure Travelers." Journal of Travel Research 32: 2-6.

Glaser, Barney G., and Anselm Strauss. 1967. The Discovery of Grounded Theory: Strategy for Qualitative Research. New York: Aldine De Gruyter.

Gluck, Frederick W. 1982. "Strategic Management for Competitive Advantage." Journal of Business Strategy 2(3):9-21.

Goeldner, Charles R., and J.R. Brent Ritchie. 2003. Tourism: Principles, Practices, Philosophies. 9th ed. Hoboken, N.J.: Wiley.

Golafshani, Nahid. 2003. "Understanding Reliability and Validity in Qualitative Research." The Qualitative Report 8(4):597-607.

Golding, Peter, and Graham Murdock. 2001. "Digital Divide: Communications Policy and its Contradictions." New Economy 8(2):110–15.

Goldsmith, Ronald E., and Stephen Litvin. 1999. "Heavy Users of Travel Agents: A Segmentation Analysis of Vacation Travelers." Journal of Travel Research 38:127.

Goossens, Cees F. 1994. "External Information Search: Effects of Tour Brochures with Experiential Information." In Economy Psychology of Travel and Tourism, edited by John Crotts, and W. Fred Van Raaij, 89-107. New York: The Haworth Press.

—. 2000. "Tourism Information and Pleasure Motivation." Annals of Tourism Research 27(2):301-321.

Gordijn, Jaap, and Hans Akkermans. 2001. "Designing and Evaluating E-business Models." IEEE Intelligent Systems 16(4):11-17.

Govers, Robert F., Frank M. Go, and Kumar Kuldeep. 2007. "Promoting Tourism Destination Image." Journal of Travel Research 46(1):15-23.

Govers, Robert F. 2004. "Projected Destination Image on the World Wide Web: Content Analysis of Online Textual Representations of Dubai." Conference Proceedings for Information and Communication Technologies in Tourism.

Graeupl, Alice, and Scott McCabe. 2003. "Use of the Internet by 'Silver Surfers' for Travel and Tourism Information Use and Decision-making: A Conceptual Review and Discussion of Findings." In Information and Communication Technologies in Tourism, edited by Andrew Frew, Martin Hitz, and Peter O'Connor, 144-155. New York: Springer.

Grandon, Elizabeth, and J. Michael Pearson. 2004. "Electronic Commerce Adoption: An Empirical Study of Small and Medium US Businesses." Information and Management 42:197-216.

Grant, Robert M. 1995. Contemporary Strategy Analysis. Oxford: Blackwell.

Gray, Colin. 2006. "Stage Models of ICT Adoption in Small Firms." In Impact of E-commerce on Consumers and Small Firms, edited by Salvatore Zappala, and Colin Gray, 3-20. Aldershot: Ashgate.

Greene, Jennifer C., and Valerie J. Caracelli. 2003. "Making Paradigmatic Sense of Mixed Methods Practice." In Handbook of Mixed Methods in Social and Behavioral Research, edited by Charles Teddlie, and Abbas Tashakkori, 91-110. London: SAGE Publications.

Greenwood, Royston, David Deephouse, and Stan Li. 2007. "Ownership and Performance of Professional Service Firms." Organization Studies 28(2):219-238.

Greenwood, Royston, and Laura Empson. 2003. "The Professional Partnership: Relic or Exemplary Form of Governance?" Organization Studies 24(6):909–33.

Grennes, Thomas. 1997. Introduction to Scientific Theory and Method. Oslo: Tano Aschehoug.

Grimm, Laurence G., and Paul R. Yarnold. 1995. Reading and Understanding Multivariate Statistics. Washington, D.C.: American Psychological Association.

Grönroos, Christian. 1993. "Toward a Third Phase in Service Quality Research." In Advances in Services Marketing and Management, edited by Teresa A. Swartz, David A. Bowen, and Stephen W. Brown, 2:49-64. Greenwich, CT: JAI Press.

Grossmann, Wilfried, and Marcus Hudec. 2003. "Web Usage Mining in Ecommerce." International Journal of eBusiness (Special Issue EC3).

Guba, Egon G. 1987. "Naturalistic Evaluation." New Directions for Program Evaluation 34:23-43.

Guillén, Mauro, and Sandra Suárez. 2005. "Explaining the Global Digital Divide: Economic, Political and Sociological Drivers of Cross-national Internet Use." Social Forces 84(2):681-708.

Gunkel, David. 2003. "Second Thoughts: Toward a Critique of the Digital Divide." New Media and Society 5(4):499–522.

Gursoy, Dogan. 2003. "Prior Product Knowledge and its Influence on the Traveler's Information Search Behaviour." Journal of Hospitality and Leisure Marketing 10(3/4):113-130.

Gursoy, Dogan, and Joseph Chen. 2000. "Competitive Analysis of Cross Cultural Information Search Behavior." Tourism Management 21:583-590.

Gursoy, Dogan, and Ken W. McCleary. 2004. "An Integrative Model of Tourists' Information Search Behavior." Annals of Tourism Research 31(2):353-373.

Gyamfi, Alexander. 2005. "Closing the Digital Divide in Sub-Saharan Africa: Meeting the Challenges of the Information Age." Information Development 21(1):22-30.

Hanssler, Karl Heinz. 2001. Management in der Hotellerie und Gastronomie. München: Oldenburg Verla.

Haertsch, Patrick. 2000. Wettbewerbstrategien für Electronic Commerce: Eine Kritische Überprüfung Klassischer Strategiekonzepte. Lohmar; Köln, Josef Eul Verlag.

Hair, Joseph F., Rolph E. Anderson, Ronald L. Tatham, and William Black. 1998. Multivariate Data Analysis. 5th edition. New Jersey: Prentice-Hall.

Hamel, Gary, and C. K. Prahalad. 1994. Competing for the Future. Boston: Harvard Business School Press.

Hamlin, Robert, and Jenny Sawyer. 2007. "Developing Effective Leadership Behaviours: The Value of Evidence-based Management." Business Leadership Review 4 (4): 1-16.

Haralambos, Michael, and Martin Holborn. 1980. Sociology Themes and Perspectives. Collins Educational.

Harland, Christine M. 1996. "Supply Chain Management, Purchasing and Supply Management, Logistics, Vertical Integration, Materials Management and Supply Chain Dynamics." In Blackwell Encyclopedic Dictionary of Operations Management, edited by Nigel Slack. UK: Blackwell.

Harrison, Graeme, and Jill McKinnon. 1999. "Cross-cultural Research in Management Control Systems Design: A Review of the Current State." Accounting, Organizations and Society 24:483–506.

Hatton, Gerry. 1994. "Designing a Warehouse or Distribution Centre." In The Gower Handbook of Logistics and Distribution Management, edited by John Gattorna, 175-193. 4th ed. Aldershot, England: Gower Publishing.

Hatton, Mike. 2004. "Redefining the Relationships: The Future of Travel Agencies and the Global Agency Contract in a Changing Distribution System." Journal of Vacation Marketing 10:101.

Hedna. 2002. "Despite Challenges in 2001: HEDNA Reports that GDS Bookings for Hotels Remain Strong." London: Hedna Press Release. Accessed January 2, 2010. www.hedna.org.

Heeks, R. 2000. "Analysing E-commerce for Development." Accessed April 7, 2009. http://www.sed.manchester.ac.uk/idpm/research/publications/wp/di/short/di_sp04.pdf

Heinich, Nathalie. 2010. "What does 'Sociology of Culture' Mean?: Notes on a Few Trans-cultural Misunderstandings." Cultural Sociology 4:257-265.

Heung, Vincent C. S. 2003. "Barriers to Implementing E-commerce in the Travel Industry: A Practical Perspective." International Journal of Hospitality Management 22(1):111-118.

Hitt, Michael A., R. Duane Ireland, and Robert E. Hoskisson. 2001. Strategic Management: Competitiveness and Globalization. 4th ed. Mason, OH: South Western Publishing.

HoCo. 2000. Worldwide Hotel Industry Study, 1996 – 2000. London: Horwath Consulting.

Hoffman, Donna, and Thomas P. Novak. 1996. "Marketing in Hypermedia Computer-mediated Environments: Conceptual Foundations." Journal of Marketing 60:50-68.

Hoffman, Eric, and Paul Roman. 1984. "Information Diffusion in the Implementation of Innovation Process." Communication Research 11(1):117-140.

Hofstede, Geert. 1980. Culture's Consequences: International Differences in Work-related Values. Beverly Hills, CA: Sage.

—. 1991. Cultures and Organizations: Software of the Mind. Maidenhead: McGraw-Hill.

Hofstede, Geert, and Michael H. Bond. 1984. "Hofstede's Culture Dimensions: An Independent Validation Using Rokeach's Value Survey." Journal of Cross-cultural Psychology 15:417-433.

Hollander, Edwin P. 1993. "Legitimacy, Power and Influence: A Perspective on Relational Features of Leadership." In Leadership Theory and Research: Perspectives and Directions, edited by Martin M. Chemers, and Roya Ayman, 29-48. San Diego: Academic Press.

Holloway, J. Christopher. 1994. The Business of Tourism. London: Pitman Publishing.

Hooff, Bart van Den, Jasper Groot, and Sander de Jonge. 2005. "Situational Influences on the Use of Communication Technologies: A Meta-analysis and Exploratory Study." Journal of Business Communication 4(2):4-27.

Houghton, Kathryn, and Heidi Winklehofer. 2004. "The Effect of Website and E-commerce Adoption on the Relationship between SMEs and their Export Intermediaries." International Small Business Journal 22:369.

Howard, John A., and Jagdish N. Seth. 1969. The Theory of Buyer Behaviour. New York: John Wiley.

Howe, Kenneth. 1988. "Against the Quantitative-Qualitative Incompatibility Thesis." Educational Researcher 17(8):10-16.

Hsieh, Sheauhsing, and Joseph T. O'Leary. 1993. "Communication Channels to Segment Pleasure Travelers." In Communication and Channel Systems in Tourism Marketing, edited by Muzaffa Uysal, and Daniel Fesenmaier, 57-75. Gloucestershire: The Hawthorn Press.

Huber, George, and Reuben McDaniel. 1986. "The Decision-making Paradigm of Organizational Design." Management Science 32(5):572-589.

Hyde, Kenneth F. 2000. "A Hedonic Perspective on Independent Vacation Planning, Decision-making and Behaviour." In Consumer Psychology of Tourism, Hospitality and Leisure, edited by Arch G. Woodside, 177-191. New York: CABI.

Inkpen, Gary. 1998. Information Technology for Travel and Tourism. Edinburgh: Addison Wesley.

James, Jeffrey. 2004. "The Global Digital Divide in the Internet: Developed Countries Constructs and Third World Realities." Journal of Information Science 31(2):114–123.

Jang, SooCheong. 2004. "The Past, Present, and Future Research of Online Information Search." Journal of Travel and Tourism Marketing 17(2/3):41-47.

Jaskyte, Kristina, and Kisieliene Audrone. 2006. "Organizational Innovation: Comparison of Nonprofit Human-service Organizations in Lithuania and the United States." International Social Work 49(2):165–176.

Jamaica Gleaner. 2000. "Travel Agents to Boycott AA." Accessed November 14, 2010.
http://jamaica-gleaner.com/gleaner/20000111/f1.html

—. 2011. "Tourism Minister Targets Illegal Travel Agents." Accessed July 27, 2011.
http://jamaica-gleaner.com/gleaner/20110724/business/business7.html

Jeng, Jiann-Min, and Daniel R. Fesenmaier. 1999. Understanding Tourist Information Search Patterns: Insight for Developing Decision Appliances. TTRA Annual Conference.

Jenkins, Roger L. 1978. "Family Vacation Decision-making." Journal of Travel Research 16(4): 2-7.

Jennings, Gayle. 2001. Tourism Research. Milton, Australia: John Wiley & Sons.

Jung, Timothy H., and Richard Butler. 2000. "The Measurement of the Marketing Effectiveness of the Internet in the Tourism and Hospitality Industry." In Information and Communication Technologies in Tourism 2000: Proceedings of the International Conference in Barcelona, Spain, edited by Daniel Fesenmaier, Stefan Klein, and Dimitrios Buhalis, 460-472. Wien: Springer.

Jung, Timothy H., Panos Louviers, and Harmen Oppewal. 2002. "Channel Management Strategy in the E-commerce Environment: A Portfolio Management Approach." In Information and Communications Technologies in Tourism, edited by Karl Wober, Andrew Frew, and Martin Hitz, 17-26. New York: Springer Verlag.

Kadison, Maria. 1998. "On-line Retail Strategies." Forrester Research 15.

Karcher, Karsten. 1997. Reinventing the Package Holiday Business. Wiesbaden: Deutscher Universitaetsverlag.

Kalakoa, Ravi, and Andrew B. Whinston. 1996. Frontiers of eCommerce. Reading: Addison-Wesley.

Karamustafa, Kurtulus. 2000. "Marketing-Channel Relationships: Turkey's Resort Purveyors' Interactions with International Tour Operators." Cornell Hotel and Restaurant Administration Quarterly 41(4):21-32.

Karnes, Frances, and Jane Chauvin. 1985. Leadership Skills Inventory Forms. Scottsdale: Great Potential Press.

Keller, Kenneth. 2008. "From Here to There in Information Technology: The Complexities of Innovation." American Behavioral Scientist 51(1):97-106.

Kelman, Herbert C. 1961. "Process of Opinion Change." Public Opinion Quarterly 25:57-78.

Kemp, Stephen. 2005. "Critical Realism and the Limits of Philosophy." European Journal of Social Theory 8:171-191.

Kerr, Barbara A. 1988. Review of the Leadership Skills Inventory. Iowa City: The University of Iowa.

Kerstetter, Deborah, and Mi-Hea Cho. 2004. "Prior Knowledge, Credibility and Information Search." Annals of Tourism Research 31(4):961-985.

Kim, Eonsoo, Nam Dae-Il, and J. L. Stimpert. 2004. "The Applicability of Porter's Generic Strategies in the Digital Age: Assumptions,

Conjectures, and Suggestions." Journal of Management 30(5):569–589.

King, William R. 1998. "IT Enhanced Productivity and Profitability." Information System Management 15:64 -68.

Kingsley, Ilana, and Daniel R. Fesenmaier. 1995. "Travel Information Kiosks: An Emerging Communications Channel for the Tourism Industry." Journal of Travel and Tourism Marketing 4(1):57-70.

Kirkman, Geoffrey, Peter Cornelius, Jeffrey Sachs, and Klaus Schwab. 2002. Global Information Technology Report 2001-2002. Oxford: Oxford University Press.

Klaus, Peter G. 1985. "Quality Epiphenomenon: The Conceptual Understanding of Quality in Face-to-face Service Encounters." In The Service Encounter: Managing Employee / Customer Interaction in Service Businesses, edited by John A. Czepiel, Michael R. Solomon, and Carol F. Surprenant, 17–33. Lexington, Mass: Lexington Books.

Klein, Stefan. 1996. Interorganisationssysteme und Unternehmensnetzwerke. St. Gallen: Universität St. Gallen

Klein, Stefan, and S. Güler. 1999. Hotel Reservation Systems on the Internet: Custom Design vs. Standard Software. Innsbruck: Springer Verlag.

Klein, Stefan, and Beat Schmid. 1996. Information and Communication Technologies in Tourism. Innsbruck, Springer Verlag.

Klein, Barbara D. 2001. "User Perceptions of Data Quality: Internet and Traditional Text Sources." The Journal of Computer Information Systems 41(4):9-15.

Klein, Stefan. 2002. Web Impact of the Distribution Structure for Flight Tickets. Innsbruck: Springer Verlag.

Kotler, Philip, and Gary Armstrong. 2008. Principles of Marketing. 12th ed. New Jersey: Pearson-Prentice Hall.

Kouzes, James M., and Barry Posner. 1987. The Leadership Challenge: How to Keep Getting Extraordinary Things Done in Organizations. San Francisco, CA: Jossey Bass.

Kowalewski, Oskar, Oleksandr Talavera, and Ivan Stetsyuk. 2010. "Influence of Family Involvement in Management and Ownership on Firm Performance: Evidence from Poland." Family Business Review 23(1):45–59.

Kraajenbrink, Jeroen, J. C. Spender, and Aard Groen. 2010. "The Resource-based View: A Review and Assessment of its Critiques." Journal of Management 36: 349.

Krippendorff, Klaus. 1989. Shannon, Claude. In International Encyclopedia of Communication, edited by Erik Barnouw, 59-61. New York: Oxford University Press.

Kuan, Kevin K. Y., and Patrick Chau. 2001. "A Perception-based Model for EDI Adoption in Small Businesses Using Technology-Organization-Environment Framework." Information and Management 38:507-521.

Kuhnert, Karl, and Philip Lewis. 1987. "Transactional and Transformational Leadership: A Constructive/Developmental Analysis." The Academy of Management Review 12(4):648-657.

Kwon, Winston, Ian Clarke, and Ruth Wodak. 2009. "Organizational Decision-making, Discourse, and Power: Integrating across Contexts and Scales." Discourse and Communication 3:273-302.

Lang, Tania. 2000. "The Effect of the Internet on Travel Consumer Purchasing Behaviour and Implications for Travel Agencies." Journal of Vacation Marketing 6: 368.

Langley, Ann, Henry Mintzberg, Patricia Pitcher, Elizabeth Posada, and Jan Saint-Macary. 1995. "Opening Up Decision Making: The View from the Black Stool." Organisation Science 6(3): 260-279.

Larsen, Tor, and Eugene McGuire. 1998. Information Systems Innovation and Diffusion: Issues and Directions. Hershey, PA: Idea Group Publishing.

Lash, Scott. 1994. "Reflexivity and its Doubles: Structure, Aesthetics, Community." In Reflexive Modernization, edited by Ulrich Beck, Anthony Giddens, and Scott Lash, 110-173. Cambridge: Polity Press.

Lasswell, Harold D. 1948. "The Structure and Function of Communication in Society." In The Communication of Ideas: A Series of Addresses, edited by Lyman Bryson, 215-228. New York: Harper and Brothers.

Latzer, Michael. 2009. "Information and Communication Technology Innovations: Radical and Disruptive?" New Media Society 11(4):599-619.

Law, Rob, and Rita Leung. 2000. "A Study of Airlines' Online Reservation Services on the Internet." Journal of Travel Research 39:202.

Law, Rob, Kenith Leung, and James Wong. 2004. "The Impact of the Internet on Travel Agencies." International Journal of Contemporary Hospitality Management 16(2):100-107.

Laws, Eric. 2001. "Distribution Channel Analysis for Leisure Travel." In Tourism Distribution Channels: Practices, Issues and Transformations, edited by Dimitrios Buhalis, and Eric Laws, 53-72. London: Continuum.

Laws, Eric, and Dimitrios Buhalis. 2001. "Agendas for Future Research." In Tourism Distribution Channels: Practices, Issues and Transformations, edited by Eric Laws, and Dimitrios Buhalis, 371-376. London: Continuum.

Lederer, Albert, Donna Maupin, Mark Sena, and Youlong Zhuang. 1990. "The Role of Ease of Use, Usefulness and Attitude in the Prediction of World Wide Web Usage." Proceedings of the Hawaii International Conference on Systems Science (HICSS)32:195-204.

Lefebvrea, Louis, Elisabeth Lefebvrea, Elie Elia, and Harold Boeck. 2005. "Exploring B-to-B e-Commerce Adoption Trajectories in Manufacturing SMEs." Technovation 25:1443-1456.

Lee, Steven W. 1988. Review of the Leadership Skills Inventory. Terre Haute, IN: Indiana State University.

Legris, Paul, John Ingham, and Pierre Collerette. 2003. "Why Do People Use Information Technology? A Critical Review of the Technology Acceptance Model." Information and Management 40:191-204.

Levy, Margi, and Philip Powell. 2002. "SME Internet Adoption: Towards a Transporter Model." In 15th Bled Electronic Commerce Conference eReality: Constructing the eEconomy, 507-521. Bled, Slovenia.

Lewis, Barbara, and Pamela McCann. 2004. "Service Failure and Recovery: Evidence from the Hotel Industry." International Journal of Contemporary Hospitality Management 16(1):6-17.

Lewis, Gordon K. 2004. The Growth of the Modern West Indies. Kingston: Ian Randle publishers.

Liikanen, Erkki. 2002. European Tourism Forum 2002. Brussels: European Commission.

Lin, Pei-Jung, Eleri Jones, and Sheena Westwood. 2009. "Perceived Risk and Risk-relievers in Online Travel Purchase Intentions." Journal of Hospitality Marketing and Management 18(8): 782-810.

Lincoln, Yvonna S., and Egon G. Guba. 1985. Naturalistic Inquiry. California: Sage Publications.

Lippert, Susan K., and Miles Davis. 2006. "A Conceptual Model Integrating Trust into Planned Change Activities to Enhance Technology Adoption Behaviour." Journal of Information Science 32(5):434-448.

Littlejohn, Stephen W. 1989. Theories of Human Communication. 3rd ed. Belmont, CA: Wadsworth.

Lo, Ada, Catherine Cheung, and Rob Law. 2002. "Information Search Behavior of Hong Kong's Inbound Travelers: A Comparison of Business and Leisure Travelers." Journal of Travel and Tourism Marketing 13(3):61-81.

Lo, Ada, Catherine Cheung, and Rob Law. 2004. "Information Search Behavior of Mainland Chinese Air Travelers to Hong Kong." Journal of Travel and Tourism Marketing 16(1):43-51.

Looney, Robert E. 1987. The Jamaican Economy in the 1980s: Economic Decline and Structural Adjustment. Boulder: Westview Press.

Lord Marshall. 2000. "On Line or On the Heap?" Journal of Vacation Marketing 6(4):292-298.

Loudon, David L., and Albert J. Della Bitta. 1993. Consumer Behaviour: Concepts and Applications. 4th ed. New York: McGraw-Hill.

Lumsdon, Les M., and Jonathan S. Swift. 1999. "The Role of Tour Operators in South America: Argentina, Chile, Paraguay and Uraguay." International Journal of Tourism Research 4(1):429-439.

Luo, Man, Ruomei Feng, and Liping A. Cai. 2004. "Information Search Behaviour and Tourist Characteristics: The Internet Vis-a-Vis other Information Sources." Journal of Travel and Tourism Marketing 17(2/3):15-25.

Lynskey, Michael. 2004. "Determinants of Innovative Activity in Japanese Technology-based Start-up Firms." International Small Business Journal 22(2):159–196.

McLean, Laird D. 2005. "Organizational Culture's Influence on Creativity and Innovation: A Review of the Literature and Implications for Human Resource Development." Advances in Developing Human Resources 7(2):226-246.

Mckay, Judith, Peter Marshall, and Adi Prananto. 2000. "Stages of Maturity for e-Business: The SOG-emodel." Accessed December 3, 2009 http://www.pacis-net.org/file/2000/29-43.pdf

Mahoney, Joseph T., and J. Rajendran Pandian. 1992. "The Resource-based View Within the Conversation of Strategic Management." Strategic Management Journal 13:363-380.

Mandell, Barbara, and Shilpa Pherwani. 2003. "Relationship between Emotional Intelligence and Transformational Leadership Style: A Gender Comparison." Journal of Business and Psychology, 17(3):387-404.

Manfredo, Michael J. 1989. "An Investigation of the Basis for External Information Search in Recreation and Tourism. Leisure Sciences 11:29-45.

Mansfeld, Yoel. 1992. "From Motivation to Travel." Annals of Tourism Research 19:399-419.

Mansell, Robin. 2002. "From Digital Divides to Digital Entitlements in Knowledge Societies." Current Sociology 50(3):407-426.

Marcussen, Carl H. 2002. "Trends in European Internet Distribution: Of Travel and Tourism Services." Accessed April 12, 2010. http://www.crt.dk/uk/staff/chm/trends.htm

Markwell, Kevin W. 1997. "Dimensions of Photography in a Nature-based Tour." Annals of Tourism Research 24:131-155.

Marler, Janet H., Xiaoya Liang., and James Hamilton Dulebhon. 2006. "Training and Effective Employee Information Technology Use." Journal of Management 32(5):721-743.

Marshall, Catherine, and Gretchen B. Rossman. 1995. Designing Qualitative Research. California: Sage Publications Inc.

Maser, Birgit, and Klaus Weiermair. 1998. "Travel Decision-making: From the Vantage Point of Perceived Risk and Information Preferences." Journal of Travel and Tourism Marketing 7(4): 107-121.

Masschelein, Lucas, and Alain Van Buyten. 2002. Structure, Performance and Competitiveness of European Tourism and its Enterprises. Price Waterhouse Coopers.

Mathieson, Alister, and Geoffrey Wall. 1982. Tourism: Economic, Physical and Social Impacts. London: Longman.

Maurer, Christian, and Veronika Lutz. 2011. "The Impact of Digital Divide on Global Tourism: Strategic Implications of Overcoming Communication Gaps caused by Digital Inequalities." In Information and Communication Technologies in Tourism 2011 Enter Proceedings, edited by Rob Law, Matthias Fuchs, and Francesco Ricci, 265-277. New York: Springer Wein.

Maxwell, Joseph A., and Diane M. Loomis. 2003. "Mixed Methods Design: An Alternative Approach." In Handbook of Mixed Methods in Social and Behavioral Research, edited by Charles Teddlie, and Abbas Tashakkori, 241-272. London: SAGE Publications.

May, Christopher. 2002. Information Society: A Sceptical View. Cambridge: Polity Press.

May, Paul. 2001. Mobile Commerce: Opportunities, Applications, and Technologies of Mobile Business. Cambridge: Cambridge University Press.

Maykut, Pamela. 1994. Beginning Qualitative Research: A Philosophical and Practical Guide. Washington, D.C.: The Falmer Press.

McIntosh, Alison. 1998. "Mixing Methods: Putting the Tourist at the Forefront of Tourism Research." Tourism Analysis 3:121-127.

McQuail, Denis, and Swen Windahl. 1981. Communication Models - For the Study of Mass Communications. London: Longman.

McSweeney, Brendan. 2002. "Hofstede's Model of National Cultural Differences and the Consequences: A Triumph of Faith – A Failure of Analysis." Human Relations 55:89–118.

Meditz, Sandra W., and Dennis Hanratty. 1987. "Caribbean Islands: A Country Study." Washington: GPO for the Library of Congress. Accessed March 14, 2011. http://countrystudies.us/caribbean-islands/24.htm

Menard, Scott. 1995. Applied Logistic Regression Analysis. Thousand Oaks, CA: Sage Publications.

Mercer, Jonathan. 2005. "Rationality and Psychology in International Politics." International Organization 59 (1):77-106

Mill, Robert C., and Alastair Morrison. 1992. The Tourism System: An Introductory Text. 2nd ed. Englewood Cliffs, N.J.: Prentice Hall.

Minghetti, Valeria, and Dimitrios Buhalis. 2010. "Digital Divide in Tourism." Journal of Travel Research 49(3):267-281.

Mitchell, Susan. 1998. American Generations: Who They Are, How They Live, What They Think. 2nd ed. Ithaca, NY: New Strategist.

Mochrie, Robert, Laura Galloway, and David Deakins. 2006. "The Value of Internet Forums to Small Rural Businesses." In Impact of e-Commerce on Consumers and Small Firms, edited by Salvatore Zappala, and Colin Gray, 81-100. Aldershot: Ashgate.

Moersch, Christopher. 1995. "Levels of Technology Implementation (loti): A Framework for Measuring Classroom Technology Use." Accessed February 5, 2010.
http://loticonnection.com/pfd/LoTiFrameworkNov95.pdf.

Moital, Miguel, Roger Vaughan, and Jonathan Edwards. 2009. "Using Involvement for Segmenting the Adoption of E-commerce in Travel." Service Industries Journal 29(5):723-739.

Money, R. Bruce, and John C. Crotts. 2003. "The Effect of Uncertainty Avoidance on Information Search, Planning, and Purchases of International Travel Vacations." Tourism Management 24: 191-202.

Montgomery, Cynthia A., and Sam Harihan. 1991. "Diversified Expansion by Large Established Firms." Journal of Economic Behaviour 52(1):71-89.

Moorthy, Sridhar, Brian T. Ratchford, and Debabrata Talukdar. 1997. "Consumer Information Search Revisited: Theory and Empirical Analysis." Journal of Consumer Research 23:263-277.

Morgan, Nigel, Annette Pritchard, and Sara Abbott. 2001. "Consumer, Travel and Technology: A Bright Future for the Web or Television Shopping?" Journal of Vacation Marketing 7(2):110-124.

Morrison, Alastair M. 2002. Hospitality and Travel Marketing. 3rd ed. Albany, Delmar Cengage Learning.

Morrison, Alastair. 2002. "Tourism, Hospitality and Leisure Journals." Accessed April 7, 2010. http://omni.cc.purdue.edu/~alltson/journals.htm

Morrison, Alastair M., Juline E. Mills, Suttichai Chuvessiriporn, and Joseph A. Ismail. 2002. "Where are We Now?: An Initial Analysis of Web-based Marketing Issues Affecting Travel and Tourism." In Information and Communications Technologies in Tourism 2002, edited by Karl Wober, Andrew Frew, and Martin Hitz, 375-386. New York: Springer Verlag.

Morrison, Alison. 1994. "Marketing Strategic Alliances: The Small Hotel Firm." International Journal of Contemporary Hospitality Management 6(3):25-30.

Morrison, Alison, Michael Rimmington, and Clare Williams. 1999. Entrepreneurship in the Hospitality, Tourism and Leisure Industries. Oxford: Butterworth-Heinemann.

Morrison, Margaret. 2002. Using Qualitative Research in Advertising: Strategies, Techniques, and Applications. California: Sage Publications.

Morse, Janice M. 2003. "Principles of Mixed Methods and Multimethod Research Design." In Handbook of Mixed Methods in Social and Behavioral Research, edited by Charles Teddlie, and Abbas Tashakkori, 189-208. London: SAGE Publications.

Morton, Michael Scott. 1991. The Corporation of the 1990s - Information Technology and Organizational Transformation. New York: Oxford University Press.

Mosaic Group. 1998. "The Global Diffusion of the Internet Project: An Initial Inductive Study." Accessed January 16, 2010. www.agsd.com/gdi97/gdi97.html

Moutinho, Luiz. 1987. "Consumer Behaviour in Tourism." European Journal of Marketing 21:5-44.

Mowery, David C., Joanne E. Oxley, and Brian Silverman. 1998. "Technological Overlap and Interfirm Cooperation: Implications for the Resource-based View of the Firm." Research Policy 27: 507-523.

Munro, Barbara H. 2001. Statistical Methods for Health Care Research. 4th ed. Philadelphia: Lippincott.

Murphy, William H. 1999. "Hofstede's National Culture as a Guide for Sales Practices across Countries: The Case of a MNC's Sales Practices in Australia and New Zealand." Australian Journal of Management 24(1):37-58.

Nelson, Richard R., and Sidney Witter. 1982. An Evolutionary Theory of Economic Behaviour and Capabilities. Cambridge: Harvard University Press, 195-307.

Neuman, W. Lawrence. 2006. Social Research Methods: Qualitative and Quantitative Approaches. 6th ed. Boston: Allyn and Bacon.

Nolan, Sidney D. 1976. "Tourists' Use and Evaluation of Travel Information Sources: Summary and Conclusions." Journal of Travel Research14(2):6-8.

Norris, Pipps. 2000. "Digital Divide?: Civic Engagement, Information Poverty and the Internet in Democratic societies." Accessed November 9, 2010. web.archive.org/web/20010217154551/http://ksghome.harvard.edu/~.pnorris.sho renstein.ksg/book1.htm

Norum, Pamela S. 2008. "Student Internet Purchases." Family and Consumer Sciences Research Journal 36:373-388.

NUA. 2003. Internet Demographics.

Nysveen, Herbjørn, and Maria Lexhagen. 2002. "Reduction of Perceived Risk through On-line Value-added Services Offered on Tourism Business Web-sites." In Information and Communications Technologies in Tourism 2002, edited by Karl Wober, Andrew Frew, and Martin Hitz, 488-496. New York: Springer Verlag.

O'Connor, Peter. 2000. Evaluating Electronic Channels of Distribution in the Hotel Sector. Barcelona: Springer Verlag.

—. 2001. "The Changing Face of Hotel Electronic Distribution." Travel and Tourism Analyst 5:61-78.

—. 2002. An Analysis of the Online Pricing Strategies of the International Hotel Chains. Innsbruck, Springer Verlag.

O'Connor, Peter, Dimitrios Buhalis, and Andrew Frew. 2001. "The Transformation of Tourism Distribution Channels through Information Technology." In Tourism Distribution Channels: Practices, Issues and Transformations, edited by Dimitrios Buhalis, and Eric Laws, 332-350. New York: Continuum.

OECD. 2001. "Organization for Economic Co-operation and Development: Measuring Electronic Commerce." OECD 1997. Accessed December 4, 2010. http://www.oecd.org/dsti/sti/it/ec/prod/e_97-185-htm

Öörni, Anssi. 2004. "Consumer Objectives and the Amount of Search in Electronic Travel and Tourism Markets." Journal of Travel and Tourism Marketing 17(2/3):3-14.

Opodo. 2003. www.opodo.com.

Opperman, Martin. 1999. "Databased Marketing by Travel Agencies." Journal of Travel Research 37(3):231-237.

Orbitz. 2003. www.orbitz.com.

O'Reilly, Charles A. 1982. "Variations in Decision Makers' Use of Information Sources: The Impact of Quality and Accessibility of Information." Academy of Management Journal 25(4): 756-771.

Oz, Effy. 2009. Managing Information Systems. 6th ed. Australia: Cengage Learning.

Page, Stephen. 2005. Transport and Tourism: Global Perspectives. 2nd ed. Harlow, England: Pearson, Prentice Hall.

Page, Stephen, and Don Getz.1997. The Business of Rural Tourism: International perspectives. London: Thomson Business Press.

Palmer, Adrian, and Patrick McCole. 1999. "The Virtual Re-intermediation of Travel Services: A Conceptual Framework and Empirical Investigation." Journal of Vacation Marketing 6(1):33-47.

Pan, Bing, and Daniel Fesenmaier. 2001. "Exploring Caller Dialogue: Analyzing Directed Information Search for Tourism Products." In Information and Communication Technologies in Tourism 2001: Proceedings of the International Conference in Montreal, Canada, 2001, edited by Pauline Sheldon, Karl Wober, and Daniel Fesenmaier, 244-250. Wien, Austria: Springer.

Pan, Bing, and Daniel Fesenmaier. 2003. "Travel Information Search on the Internet: A Preliminary Analysis." In Information and Communication Technologies in Tourism 2003, edited by Andrew Frew, Martin Hitz, and Peter O'Connor, 242-251.

Parker, Marilyn. 1983. Information Economics. London, Prentice-Hall.

Passmann, Christoph, Wolfgang Pipperger, and Walter Schertler. 1997. "How to Assess WWW- Applications for Tourism Information Systems from the End-User Perspective- Methodological Design and Empirical Evidence." In Information and Communication Technologies in Tourism: Proceedings of the International Conference in Edinburgh, Scotland, edited by A. Min Tjoa, 208-219. Wien: Springer.

Patricia, Dodu. 2008. "The Internet, Threat or Tool for Travel Agencies?" Accessed April 9, 2010. http://steconomice.uoradea.ro/anale/volume/2008/v2-economy-and business-administration/017.pdf

Patton, Michael Quinn. 2002. Qualitative Evaluation and Research Methods. 3rd ed. Thousand Oaks, CA: Sage Publications.

Pawar, Badrinarayan S. 2003. "Central Conceptual Issues in Transformational Leadership Research." Leadership and Organization Development Journal 24(7):397-406.

Pawar, Badrinarayan S., and Kenneth Eastman. 1997. "The Nature and Implications of Contextual Influences on Transformational Leadership: A Conceptual Examination." Academy Management Review 22: 80-109.

Pearce, Douglas G. 2009. "Beyond Tiers: A Network Approach to Tourism Distribution." Tourism Analysis 13:517-530.

Pennington-Gray, Lori, and Christine Vogt. 2003. "Examining Welcome Center Visitors' Travel and Information Behaviors: Does Location of Centres or Residency Matter?" Journal of Travel Research 41:272-280.

Penrose, Edith T. 1959. The Theory of the Growth of the Firm. New York: Wiley.

Perdue, Richard R. 1985. "Segmenting State Information Inquirers by Timing of Destination Decision and Previous Experience." Journal of Travel Research 23:6-11.

—. 1993. "External Information Search in Marine Recreational Fishing." Leisure Sciences 15:169-187.

Perry, Chad A. 1998. Structured Approach to Presenting Theses. Accessed February 25, 2009.
http://www.scu.edu.au/schools/gcm/ar/art/cperry.html

Persson, Goran. 1978. "Organisation Design Strategies for Business Logistics." International Journal of Physical Distribution and Materials Management 8:(6).

Peteraf, Margaret A. 1993. "The Cornerstones of Competitive Advantage: A Resource-based View." Strategic Management Journal 14:179-191.

Peterson, Karen Ida. 1994. "Qualitative Research Methods for the Travel and Tourism Industry." In Travel, Tourism, and Hospitality Research: A Handbook for Managers, edited by J. R. Brent Ritchie, and Charles R. Goeldner, 487-489. 2nd edition. London: John Wiley and Sons.

Peterson, Suzanne J., Kristin Byron Walumbwa, and Jason Myrowitz. 2009. "CEO Positive Psychological Traits, Transformational Leadership, and Firm Performance in High-technology Start-up and Established Firms." Journal of Management 35:348-368.

Pettigrew, Andrew M. 1987. "Context and Action in the Transformation of Firms." Journal of Management Studies 24:649-70.

PhoCusWright. 2003. Consumer Trends: Evolution of the Online Traveler. Sherman: PhoCusWright 2.

Poon, Auliana. 1993. Tourism, Technology and Competitive Strategies. Wallingford: CAB International.

—. 2001. "The Future of Travel Agents." Travel & Tourism Analyst 3:57-80.

Poon, Simpson, and Mathew Joseph. 2001. "A Preliminary Study of Product Nature and Electronic Commerce." Marketing Intelligence & Planning 19(7):493-499.

Porter, Michael. 1980. Competitive Strategy. New York: The Free Press.

—. 1985. Competitive Advantage. New York: The Free Press.

—. 1990. The Competitive Advantage of Nations. London: Macmillan.

—. 1995. Wettbewerbsstrategien (Competitive Strategy). New York: Campus.

—. 2001. Bewährte Strategien Werden mit dem Internet Noch Wirksamer. Harvard Business Manager 5: 64-81.

Porter, Michael, and Victor Millar. 1985. "How Information Gives You Competitive Advantage." Harvard Business Review 63(4):149-160.

Presutti, William. 2003. "Supply Chain Management and E-procurement: Creating Value-added in the Supply Chain." Industrial Marketing Management 32:219-226.

Radford, Gary P. 2005. On the Philosophy of Communication. Belmont: Thomson Wadsworth.

Raitz, Karl, and Meftah Dakhil. 1989. "A Note about Information Sources for Preferred Recreational Environments." Journal of Travel Research 27:45-49.

Rao, Subba, Glenn Metts, and Carlo Monge. 2003. "Electronic Commerce Development in Small and Medium Sized Enterprises: A Stage Model and its Implications." Business Process Management Journal 9(1): 11-32.

Rappa, Michael. 2002. "Business Models on the Web." Accessed February 7, 2010. http://digitalenterprise.org/models/models.html

Rayport, Jeffrey F. 1995. "Exploiting the Virtual Value Chain." Harvard Business Review 73(6): 75-85.

Rayport, Jeffrey, and Bernard Jaworski. 2002. Introduction to E-commerce. 2nd ed. Boston: Irwin/McGraw-Hill.

Redding, S.Gordon. 1994. "Comparative Management Theory: Jungle, Zoo or Fossil Bed?" Organization Studies 15:323-59.

Reisinger, Yvette, and Lindsay W. Turner. 2002. "Cultural Differences between Asian Tourist Markets and Australian Hosts: Part 2." Journal of Travel Research 40:385 -395.

Reve, Torger. 1990. "The Firm as a Nexus of Internal and External Contracts." In The Firm as a Nexus of Treaties, edited by Masahiko Aioki, 136-188. London: Sage.

Reynolds, Thomas J., and Jonathan Gutman. 1988. "Laddering Theory, Method, Analysis, and Interpretation." Journal of Advertising Research 28:11-31.

Ricardo, David. 1817. Principles of Political Economy and Taxation. London: J. Murray.

Ricketts, Martin. 1994. The Economies of Business Enterprise. Hempstead: Harvester Wheatsheaf.

Rideng, Arne, and Berit Grue. 2002. Foreign Visitor Survey 2002. Oslo: TØI.

Riley, Roger. 1996. "Revealing Socially Constructed Knowledge through Quasi-structured Interviews and Grounded Theory Analysis." Journal of Travel and Tourism Marketing 5(1/2):21-40.

Riley, Roger W., and Lisa Love. 1999. "The State of Qualitative Tourism Research." Annals of Tourism Research 27(1):164-187.

Rodan, Simon, and Charles Galunic. 2004. "More than Network Structure: How Knowledge Heterogeneity Influences Managerial Performance and Innovativeness." Management Journal 25 (6):541-562.

Roehl, Wesley S., and Daniel Fesenmaier. 1992. "Risk Perceptions and Pleasure Travel: An Exploratory Analysis." Journal of Travel Research 30(4):17-26.

Rogers, Everett. 1986. Communication Technology: The New Media in Society. New York: NY Free Press.

Rogers, Everett. 1962. Diffusion of Innovations. New York: The Free Press.

—. 1976. "New Product Adoption and Diffusion." Journal of Consumer Research 2: 290.

—. 1983. Diffusion of Innovations. New York: The Free Press.

—. 1994. A History of Communication Study - A Biographical Approach. New York: The Free Press.

—. 2003. Diffusion of Innovations. 5th ed. New York: The Free Press.

Rogers, Everett, and Thomas Valente. 1993. "A History of Information Theory in Communication Research." In Between Communication and Information - Information and Behavior, edited by Jorge Reina Schement, and Brent D. Ruben, 35-56. New Brunswick, NJ: Transaction.

Rossman, Gretchen B., and Bruce Wilson. 1985. "Numbers and Words: Combining Quantitative and Qualitative Methods in a Single Large-scale Evaluation Study." Evaluation Review 9(5):627.

Rushton, Alan, and John Oxley. 1993. Handbook of Logistics and Distribution Management. London: Kogan Page Limited.

Ryals, Lynette, and Andre Humphries. 2007. "Managing Key Business-to-Business Relationships: What Marketing Can Learn from Supply Chain Management." Journal of Service Research 9: 312.

Ryan, Bryce, and Neal Gross. 1943. "The Diffusion of Hybrid Seed Corn in Two Iowa Communities." Rural Sociology 8:15-24.

Saffu, Kojo, and John Walker. 2008. "Strategic Value and Electronic Commerce Adoption among Small and Medium-sized Enterprises in a Transactional Economy." Journal of Business and Industrial Marketing 23(6):395-404.

Sarros, James C., Brain K. Cooper, and Joseph Santora. 2008. "Building a Climate for Innovation through Transformational Leadership and Organizational Culture." Journal of Leadership & Organizational Studies 15: 145-158.

Sassi, Sinikka. 2005. "Cultural Differentiation or Social Segregation? Four Approaches to the Digital Divide." New Media and Society 7(5):684-700.

Schary, Philip, and Tage Larsen. 1998. Managing the Global Supply Chain. Copenhagen Business School Press.

Schiffman Leon G., and Leslie Kanuk. 2000. Consumer Behavior. Upper Saddle River, NJ: Prentice-Hall.

Schiefer, Alex. 2002. eCommerce 2001/2002. Vienna: Statistik Austria.

Schmidt, Jeffrey B., and Richard A. Spreng. 1996. "Proposed Model of External Consumer Information Search." Academy of Marketing Science Journal 24(3):246-270.

Schmitz, Joseph, and Janet Fulk. 1991. "Organizational Colleagues, Media Richness, and Electronic Mail: A Test of the Social Influence Model of Technology Use." Communication Research 18(4):487-523.

Schramm, Wilbur. 1954. "How Communication Works." In The Process and Effect of Mass Communication, edited by Wilbur Lang Schramm, 3-26. Urbana: University of Illinois Press.

Schul, Patrick, and John L. Crompton. 1983. "Search Behaviour of International Vacationers: Travel-specific Lifestyle and Sociodemographic Variables." Journal of Travel Research 21:25-31.

Schwartz, Shalom. 1992. "Universals in the Content and Structure of Values: Theoretical Advances and Empirical Tests in 20 Countries." Advances in Experimental Social Psychology 25: 1-65.

Schweda, Anika. 2004. "Independent International Traveller Evaluations of Traditional and Interactive Holiday Information Sources along Temporal and Utility Dimensions." In Information and Communication Technologies in Tourism, edited by Andrew J. Frew, 151-160. Vienna: Springer Verlag.

Seidman, Irving. 1998. Interviewing as Qualitative Research- A Guide for Researchers in Educational and Social Sciences. 2nd ed. New York: Teachers College Press.

Shailer, Gregory. 1994. "Capitalists and Entrepreneurs in Owner-managed Firms." International Small Business Journal 12:33-41.

Shannon, Claude E., and Warren Weaver. 1963. The Mathematical Theory of Communication. Urbana and Chicago: University of Illinois Press.

Sheldon, Pauline. 1997. Tourism Information Technology. New York: Cab International.

Sherlock, Philip, and Hazel Bennett. 1998. The Story of the Jamaican People. Kingston: Ian Randle Publishers.

Shin, Namchul. 2001. "Strategies for Competitive Advantage in Electronic Commerce." Journal of Electronic Commerce Research 2(4):164-171.

Simon Herbert Alexander. 1957. Models of Man. New York: John Wiley and Sons.

Simpson, Christopher. 1994. Science of Coercion: Communication Research and Psychological Warfare, 1945-1960. Oxford, England: Oxford University Press.

Sims-Shouten, Wendy, Sarah Riley, and Carla Willig. 2007. "Critical Realism in Discourse Analysis: A Presentation of a Systematic Method of Analysis Using Women's Talk of Motherhood, Childcare and Female Employment as an Example." Theory and Psychology 17: 101-124.

Singh, Niti, and R. Venkat Krishnan. 2007. "Transformational Leadership in India: Developing and Validating a New Scale." International Journal of Cross Cultural Management 7:219.

Singh, Surendra P., Ruthie G Reynolds, and Safdar Muhammad. 2001. "A Gender-based Analysis of Micro and Small Enterprises in Java, Indonesia." Journal of Small Business Management 39(2): 174-182.

Smith, Karen. 2007. "Distribution Channels for Events: Supply and Demand-side Perspectives." Journal of Vacation Marketing 13:321-338.

Snepenger, David J. 1987. "Segmenting the Vacation Market by Novelty Seeking Role." Journal of Travel Research 26:8-14.

Snepenger, David J., Kelli Meged, Mary Snelling, and Kelly Worrall. 1990. "Information Search Strategies by Destination-naive Tourists." Journal of Travel Research 29:13-16.

Snepenger, David J., and Mary Snepenger. 1993. "Market Structure Analysis of Media Selection Practices by Travel Services." In Communication and Channel Systems in Tourism Marketing, edited by Sheauhsing Hsieh, and Daniel R. Fesenmaier, 21-36. New Jersey: The Hawthorn Press.

Sparks, Colin. 2000. "The Distribution of Online Resources and the Democratic Potential of the Internet." In Cultural, Economic and

Policy Foundations for Media Openness and Diversity in East and West, edited by Jan van Cuilenburg, and Richard van der Wurff, 229-254. Amsterdam: Het Spinhuis.

Stansfield, Mark, and Kevin Grant. 2003. "An Investigation into Issues Influencing the Use of the Internet and Electronic Commerce among Small-Medium Sized Enterprises." Journal of Electronic Commerce Research 4(1):15-33.

Standing, Craig, and Thandarayan Vasudavan. 1999. "Internet Marketing Strategies Used by Travel Agencies in Australia." Journal of Vacation Marketing 6:21-32.

Statistical Institute of Jamaica. Accessed November 14, 2010. http://statinja.gov.jm/population.aspx

Straub, Evan. 2009. "Understanding Technology Adoption: Theory and Future Directions for Informal Learning." Review of Educational Research 79:625-649.

Strilka, Bernhard. 2002. Das Internet als Marketinginstrument im Tourismus - unter besonderer Berücksichtigung der österreichischen Klein- und Mittelbetriebsstruktur. Vienna: University of Business and Economics Administration.

Stone, Carl. 1973. Class, Race, and Political Behaviour in Urban Jamaica. Institute of Social and Economic Research, Mona: University of the West Indies.

—. 1985. Class, State, and Democracy in Jamaica. Kingston: Blackett Publishers.

Stonehouse, George, and Brian Snowdon. 2007. "Competitive Advantage Revisited: Michael Porter on Strategy and Competitiveness." Journal of Management Inquiry 16(3):256-273.

Stump, Rodney L., Wen Gong, and Zhan Li. 2008. "Exploring the Digital Divide in Mobile-phone Adoption Levels across Countries: Do Population Socioeconomic Traits Operate in the Same Manner as Their Individual-level Demographic Counterparts?" Journal of Macromarketing 28: 397-412.

Subramaniam, Nava, and Neal Ashkanasy. 2001. "The Effect of Organisational Culture Perceptions on the Relationship between Budgetary Participation and Managerial Job-related Outcomes." Australian Journal of Management 26(1):35-54.

Swanson, E. Burton. 1987. "Information Channel Disposition and Use." Decision Sciences 18: 131- 145.

Tabachnick, Barbara G., and Linda Fidell. 2001. Using Multivariate Statistics. 4th ed. Boston: Allyn and Bacon.

Tan, Wen Dee, and David R. Arnott. 1999. "Managerial Information Acquisition and the World Wide Web: A Theoretical Background." ISDSS '99.

Tang , Lijun, and Peidong Yang. 2011. "Symbolic Power and the Internet: The Power of a 'Horse'." Media Culture Society 33:675-691.

Tangpong, Chanchai, and Nongkran Lertpittaypoom. 2009. "The Emergence of Business-to-Customer E-commerce: New Niche Formation, Creative Destruction and Contingency Perspectives." Journal of Leadership and Organizational Studies 16(2):131-140.

Tashakkori, Abbas, and Charles Teddlie. 2003. "The Past and Future of Mixed Methods Research: From Data Triangulation to Mixed Model Designs." In Handbook of Mixed Methods in Social and Behavioral Research, edited by Charles Teddlie, and Abbas Tashakkori, 671-702. London: SAGE Publications.

Tatum, B. Charles, Richard Eberlin, Carin Kottraba, and Travis Bradberry. 2003. "Leadership, Decision-making and Organizational Justice." Management Decision 41(10):1006-1016.

Taylor, David. 1999. Global Cases in Logistics and Supply Chain Management. London: International Thomson business press.

Tchaïcha, Jane D., and Mark M. Davis. 2005. "The Impact of Culture on Technology and Business: An Interdisciplinary, Experiential Course Paradigm." Journal of Management Education 29(5):738-757.

Tebbutt, Jane. 2006. "E-Business and Small Firms in London." In Impact of E-Commerce on Consumers and Small Firms, edited by Salvatore Zappala, and Colin Gray, 63-80. Aldershot: Ashgate.

Teece, David J. 1982. "Toward an Economic Theory of the Multiproduct Firm." Journal of Economic Behaviour and Organization 3:39-63.

Teo, Thompson, Sijie Lin, and Kee-hung Lai. 2009. "Adopters and Non-adopters of E-procurement in Singapore: An Empirical Study." Omega 37(5):972-987.

Terpstra, Vern, and Kenneth David. 1991. The Cultural Environment of International Business. Cincinnati, OH: South Western Publishing.

Tetzeli, Rick. 1994. "The Internet and Your Business." Fortune 129(5):86 -96.

Theobald, Axel, and Marcus Dreyer. 2001. Online-Marktforschung. Wiesbaden: Gabler Verlag.

Thomas, Robyn, and David Dunkerley. 1999. "Careering Downwards?: Middle Management in the Downsized Organization." British Journal of Management 10(2):157-69.

Thomas, Robyn, Cynthia Hardy, and Leisa Sargent. 2011. "Managing Change: Negotiating Meaning and Power-resistance Relations." Organization Science 22(1):22-41.

Thompson, John L. 1990. Strategic Management. London: Chapman & Hall.

Thong, James Y.L., and C. S. Yap. 1995. "CEO Characteristics, Organisational Characteristics and Information Technology Adoption in Small Business." Omega-international Journal of Management Science 23 (4): 429-442.

Tierney, Patrick T. 1993. "The Influence of State Travel Information Centers on Tourist Length of Stay and Expenditure." Journal of Travel Research 31(3):28-32.

Timmers, Paul. 1998. "Business Models for Electronic Markets." Electronic Markets 8:3-8.

Tjostheim, Ingvar. 1997. "The New Network in the Norwegian Tourism Industry- Marketing and Organisational Theories as Perspectives for a Study." In Information and Communication Technologies in Tourism: Proceedings of the International Conference in Edinburgh, Scotland, 1997, edited by A. Min Tjoa, 110-118. Wien: Springer.

—. 2002. "The Internet in Competition with the Traditional Information Sources- A Study of Vacation Planning for Package vs. Non-package Travel." In Information and Communications Technologies in Tourism 2002, edited by Karl Wober, Andrew J. Frew, and Martin Hitz, 166-176. New York: Springer Verlag.

Tiscover. 2003. "Tiscover Company Profile." www.aboutiscover.com.

Tomasetto, Carlo, and Felice Carugati. 2006. "Social Influence and Diffusion of Innovations in Education." In Impact of E-Commerce on Consumers and Small Firms, edited by Salvatore Zappala, and Colin Gray, 21-34. Aldershot: Ashgate.

Toomela, Aaro. 2003. "How Should Culture Be Studied?" Culture & Psychology 9(1):35-45.

Trenholm, Sarah. 1991. Human Communication Theory. 2nd ed. Englewood Cliffs, NJ: Prentice Hall.

Trevino, Linda K., Richard L. Daft, and Robert Lengel. 1990. "Understanding Managers' Media Choices: A Symbolic Interactionist Perspective." In Organizations and Communication Technology, edited by Janet Fulk, and Charles W. Steinfield, 71-94. Newbury Park, Ca: Sage.

Trochim, William M. K. 2006. "Research Methods Knowledge Base." Accessed November 17, 2009. www.socialresearchmethods.net/kb/positivism.php

Trompenaars, Fons. 1993. Riding the Waves of Culture: Understanding Cultural Diversity in Business. London: Economist Books.

Tsai, Henry, and Zheng Gu. 2007. "Institutional Ownership and Firm Performance: Empirical Evidence from U.S. Based Publicly Traded Restaurant Firms." Journal of Hospitality and Tourism Research 31(1):19-38.

Tsao, Chiung-Wen. 2009. "Founding-family Ownership and Firm Performance: The Role of High-performance Work Systems." Family Business Review 22(4):319-332.

Tylor, Edward. 1897. Primitive Culture. New York: J.P. Putnam's Sons.

Ulijn, Jan, Dan O'Hair, Mathieu Weggeman, Gerald Ledlow, and Thomas Hall. 2000. "Innovation, Corporate Strategy, and Cultural Context: What is the Mission for International Business Communication?" Journal of Business Communication 37(3):293-317.

Um, Seoho, and John L. Crompton. 1990. "Attitude Determinants in Tourism Destination Choice." Annals of Tourism Research 17:432-448.

U.S. Department of Commerce. 2007. "Quarterly Retail E-commerce Sales: 1st Quarter 2007 U.S. Census Bureau News." Accessed March 17, 2009. www.commerce.gov

Uysal, Muzaffer, Cary D. McDonald, and Laurel Reid. 1990. "Sources of Information Used by International Visitors to U.S. Parks and Natural Areas." Journal of Park and Recreation Administration 8(1): 51-59.

Van Raaij, W. Fred. 1986. "Consumer Research on Tourism: Mental and Behavioural Constructs." Annals of Tourism Research 13:1-9.

Van Raaij, W. Fred, and Dick A. Francken. 1984. "Vacation Decisions, Activities, and Satisfactions." Annals of Tourism Research 11(1): 101-112.

Vaughan, Roger. 2008. "Conceptual Framework." Power Point Presentation, Bournemouth University. Accessed May 14, 2010. http://www.powershow.com/view/a57f7-NDM1Y/Conceptual_Framework_Professor_Roger_Vaughan_May_29th_2008_flash_ppt_presentation

Veal, Anthony James. 1997. Research Methods for Leisure and Tourism. 2nd ed. London: Pitman Publishing.

Venkatesh, Viswanath, and Fred Davis. 2000. "A Theoretical Extension of the Technology Acceptance Model: Four Longitudinal Field Studies." Management Science 46(2):186-204.

Victorino, Liana, Rohit Verma, Gerhard Plaschka, and Chekitan Dev. 2006. "Service Innovation and Customer Choices in the Hospitality Industry." Managing Service Quality 15(6):555-576.

Virolainen, Veli. 1998. "A Survey of Procurement Management Development in Industrial Companies." International Journal of Productions Economics 56:677-688.

Vishwanath, Arun, and Gerald Goldhaber. 2003. "An Examination of the Factors Contributing to Adoption Decisions among Late-diffused Technology Products." New Media and Society 5(4): 547-572.

Vogt, Christine A., and Daniel R. Fesenmaier. 1998. " Expanding the Functional Information Search Model." Annals of Tourism Research 25(3):551-578.

Vogt, Christine A., Susan I. Stewart, and Daniel R. Fesenmaier. 1998. "Communication Strategies to Reach First-time Visitors." Journal of Travel and Tourism Marketing 7(2):69-89.

Vrana, Vasiliki, and Costas Zafiropoulos. 2006. "Tourism Agents' Attitudes on Internet Adoption: An Analysis from Greece." International Journal of Contemporary Hospitality Management 18 (7):601-608.

Walle, Alf H. 1997. "Quantitative Versus Qualitative Tourism Research." Annals of Tourism Research 24(3):524-536.

Wacquant, Loic J.D. 1996. "The Rise of Advanced Marginality. Notes on its Nature and Implications." Acta Sociologica 39(2):121–39.

Waldman, David A., Bernard M. Bass, and Francis Yammarino. 1990. "Adding to Contingent Reward Behavior." Group and Organisation Studies 15:382-94.

Walters, Bruce A., Mark Kroll, and Peter Wright. 2008. "CEO Ownership and Effective Boards: Impacts on Firm Outcomes." Strategic Organization 6(3):259-283.

Warden, Stuart, and S. Tunzelana. 2004. "E-Commerce: A Critical Review of SMME Organizational Barriers in Tourism." Paper presented at the Proceedings of 6th WWW Applications Conference 2004, Rand Afrikaans University, Johannesburg, South Africa.

Wasilczuk, Julita. 2000. "Advantageous Competence of Owner/ Managers to Grow the Firm in Poland: Empirical Evidence." Journal of Small Business Management 38(2):88-94.

Weis, Hans C., and Peter Steinmetz. 2000. Martkforschung. Ludwigshafen: Kiehl Verlag.

Weiss, Michael J. 2003. "To Be About to Be." American Demographics 25(7):29-36.

Welch, Emma. 2003. "The Relationship between Ownership Structure and Performance in Listed Australian Companies." Australian Journal of Management 28(3):287-306.

Weller, Susan C., and Kimball Rommey. 1988. Systematic Data Collection. Newbury Park, California: SAGE Publications.
Wellner, Alison S. 2000. "Generational Divide." American Demographics 22(10):52-58.
Wernerfelt, Birger. 1984. "A Resource-based View of the Firm." Strategic Management Journal 5:171-180.
Werthner, Hannes. 1995a. Design Principles of Tourism Information Systems. Wien, Oldenburg.
Werthner, Hannes. ed. 1995b. Elektronisches Marketing im Tourismus. Wien: Oldenburg.
Werthner, Hannes. 1996. Design Principles of Tourism Information Systems. Innsbruck: Springer Verlag.
Werthner, Hannes, and Stefan Klein. 1999. Information Technology and Tourism: A Challenging Relationship. Vienna: Springer Verlag.
Werthner, Hannes, and Stefan Klein. 1999. "ICT and the Changing Landscape of Global Tourism Distribution." Electronic Markets 9(4):256-262.
Westwood, Robert, and David Low. 2003. "The Multicultural Muse Culture, Creativity and Innovation." International Journal of Cross Cultural Management 3(2):235-259.
Whyte, Millicent. 1983. A Short History of Education in Jamaica. London: Hodder and Stoughton.
Wicks, Bruce E., and Michael Schuett. 1993. "Using Travel Brochures to Target Frequent Travellers and "Big-spenders"." In Communication and Channel Systems in Tourism Marketing, edited by Sheauhsing Hsieh, and Daniel R. Fesenmaier, 77-89. The Hawthorn Press.
Wiertz, Caroline. 2001. "The Internet Adoption Decision in Small and Medium Enterprises (SMEs)." Accessed February 12, 2010. http://www.bonissen.de/ib/papers/Small%20Biz.pdf
Willcocks, Leslie. 2000. Moving to E-business: The Ultimate Guide to Effective E-business. London: Random House Business Book.
Williams, Densil, and Andrew Spencer. 2010 "Advertising and Tourist Arrivals: Evidence from Jamaica." Tourism and Hospitality Research 10(4):359–366.
Williamson, Dermot. 2002. "Forward from a Critique of Hofstede's Model of National Culture." Human Relations 55(11):1373-1395.
Williamson, Oliver E. 1979. "Transaction Cost-Economies: The Governance of Contractual Relations." Journal of Law and Economics 22:232-261.
Williamson, Oliver E. 1985. The Economic Institutions of Capitalism. New York: Free Press.

—. 1990. "The Firm as a Nexus of Treaties: An Introduction." In The Firm as a Nexus of Treaties, edited by Masahiko Aioki, 136-188. London: Sage.

Williamson, Oliver E., and Scott E. Masten. 1999. The Economics of Transaction Costs. Cheltenham: Edward Elgar Publishing.

Willis, Suzanne, and Bruce Tranter. 2006. "Beyond the 'Digital Divide': Internet Diffusion and Inequality in Australia." Journal of Sociology 42(1):43–59.

Wilson, Kenneth R., Jennifer S. Wallin, and Christa Reiser. 2003. "Social Stratification and the Digital Divide." Social Science Computer Review 21:133-143.

Winston, Brian. 1986. Misunderstanding Media. London: Routledge and Kegan Paul.

Wirtz, Bernd W. 2001. Electronic Business. Wiesbaden: Gabler Verlag.

Wöber, Karl. 2003. "Information Supply in Tourism Management by Marketing Decision Support Systems." Tourism Management 24:241-255.

Wöber, Karl, and Ulrike Gretzel. 2000. "Tourism Managers' Adoption of Marketing Decision Support Systems." Journal of Travel Research 39:172-181.

Wöber, Karl, and Andrew Frew. 2002. Information and Communication Technologies in Tourism 2002. Innsbruck: Springer Verlag.

Woiceshyn, Jaana. 2000. "Technology Adoption: Organizational Learning in Oil Firms." Organization Studies 21(6):1095-1118.

Wolfe, Kara, Cathy H. C. Hsu, and Soo Kang. 2004. "Buyer Characteristics among Users of Various Travel Intermediaries." Journal of Travel and Tourism Marketing 17(2/3): 51-62.

Woodside, Arch G., and Ilkka Ronkainen. 1980. "Vacation Travel Planning Segments: Self Planning Vs. Users of Motor Club and Travel Agents." Ananls of Tourism Research 7:385-393.

Woodward, Tracy. 2000. "Using Brand Awareness and Brand Image in Tourism Channels of Distribution." Journal of Vacation Marketing 6: 119-130.

WTO. 2000. World Tourism Organisation's Tourism 2020. Madrid: World Tourism Organisation.

WTTC. 2002. The Tourism Satellite Account Method of the World Travel and Tourism Council. London: WTTC.

Wynne, Clive, Pierre Berthon, Leyland Pitt, Michael Ewing, and Julie Napoli. 2001. "The Impact of the Internet on the Distribution Value Chain: The Case of the South African Tourism Industry." International Marketing Review 18(4): 420-431.

Xue, Susan. 2005. "Internet Policy and Diffusion in China, Malaysia and Singapore." Journal of Information Science 31(3): 238-250.

Yin, Robert K. 1994. Case Study Research: Design and Methods. 2nd ed. Newbury Park, CA: Sage Publications.

Ylänne-Mcewen, Virpi. 2004. "Shifting Alignment and Negotiating Sociality in Travel Agency Discourse." Discourse Studies 6:517-536.

Youn, Seounmi H. 2000. "The Dimensional Structure of Consumer Buying Impulsivity: Measurement and Validation." PhD Diss., University of Minnesota.

Yu, A. K. L., and Rob Law. 2000. The Application of eCommerce to Enhance the Competitive Advantage of Hotels in Hong Kong. Barcelona: Springer Verlag.

Yu, Liangzhi. 2006. "Understanding Information Inequality: Making Sense of the Literature of the Information and Digital Divides." Journal of Librarianship and Information Science 38(4):229-252.

Yun, Jin-Hyo, Sangmoon Park, and Mohan Avvari. 2011. "Development and Social Diffusion of Technological Innovation: Cases Based on Mobile Telecommunications in National Emergency Management." Science Technology Society 16:215-234.

Zaichkowsky, Judith L. 1985. "Measuring the Involvement Construct." Journal of Consumer Research 12(3):341-352.

Zappala, Salvatore, and Colin Gray. 2006. Impact of e-Commerce on Consumers and Small Firms. Aldershot: Ashgate.

Zappala, Salvatore, and Guido Sarchielli. 2006. "Climate for Innovation, Attitude to Internet and ICT Adoption in Small Firms." In Impact of e-Commerce on Consumers and Small Firms, edited by Salvatore Zappala, and Colin Gray, 35-50. Aldershot: Ashgate.

Zhou, Yuqiong. 2008. "Voluntary Adopters Versus Forced Adopters: Integrating the Diffusion of Innovation Theory and the Technology Acceptance Model to Study Intra-organization Adoption." New Media Society 10:475.

Zipf, George K. 1949. Human Behavior and the Principle of Least Effort. Mass: Addison-Wesley.

Zmud, Robert W. 1987. "An Empirical Investigation of the Dimensionality of the Concept of Information." Decision Sciences 9:187-195.

APPENDIX A

SUMMARY OF ONLINE ADOPTION MODELS

Moersch, (1995)	Six stages: 1. Information seeking 2. Exploration 3. Ecommerce adoption 4. Integration and Content Processing 5. Collaboration 6. Refinement
Burgess and Cooper (1998)	Three stages: 1. Electronic promotion of products 2. Customer interaction 3. Online orders, sales and payment
Allcock et al. (1999)	Four stages: 1. Threshold (computer introduction) 2. Beginner (limited networks) 3. Intermediate (static websites) 4. Advanced (email, intranet, extranet)
Earl (2000)	Six stages: 1. Homepages (external communication) 2. Internal Communication 3. Online buying and Selling 4. E-business conversion 5. E-enterprise 6. Continuous re-inventing
Heeks (2000)	Four stages: 1. Simple email use 2. Simple website use 3. Online transactions 4. Service Delivery

Mackay et al. (2000)	Six stages: 1. No online presence 2. Static online presence 3. Interactive online presence 4. Complete internet transactions 5. Integration of front and back office 6. Extended enterprise
Willcocks (2000)	Four stages: 1. Web page development 2. Transaction system development 3. Process integration 4. E-business
Wiertz (2001)	Four stages: 1. Access 2. E-procurement 3. Online promotions 4. E-sales
Daniel et al. (2002)	Four stages: 1. Developers (minimal ecommerce) 2. Communicators 3. Web presence 4. Transactors
Levy and Powell (2002)	Four stages: 1. Email and Informative websites 2. Internal and external communication 3. Business Networking 4. Electronic data interchange systems
Rayport andJaworski (2002)	Three stages: 1. Broadcast (static websites) 2. Internet interaction with customers 3. Online transactions
Rao et al. (2003)	Four stages: 1. Web presence 2. Two-way communication 3. Online transactions 4. Enterprise integration

Chan and Swatman (2004)	Four stages: 1. Ecommerce adoption 2. Centralized ecommerce 3. New technologies 4. Customer satisfaction
Beck et al. (2005)	Four stages: 1. Online advertising 2. Online sales and after sales services 3. Online procurement 4. Electronic data interchange systems
Gatautis andNeverauskas(2005)	Three stages: 1. Electronic data interchange systems 2. Centralised ecommerce 3. Global ecommerce
Lefebvrea et al.(2005)	Six stages: 1. Ecommerce non-adopters 2. Non-adopters with interest 3. Electronic content creation 4. Electronic transactions 5. Complex electronic transactions 6. Electronic collaboration
Gandhi (2006)	Four stages: 1. Online promotions 2. Online customer interaction 3. Online ordering and payment 4. Feedback and after sales services
Al-Qirim (2007)	Two stages: 1. Starters (email and passive websites) 2. Advanced level (website sales)
Chen and McQueen (2008)	Four stages: 1. Information search and email communication 2. Static online marketing websites 3. Online ordering with manual payment 4. Online transactions and invoicing

Index

D

E

F

G

H

I

U

V